T0295108

CLIMATE CHANGE AND GREENHOUSE GASES EMISSIONS

About the Authors

Dr. Pratap Bhattacharyya is ICAR-National Fellow and Principal Scientist in the Division of Crop Production in ICAR-National Rice Research Institute (NRRI), Cuttack, Odisha, India. He has 20 years of research and teaching experience on the field of climate change, GHGs emissions, environmental sciences, soil science, microbiology, soil conservation and nutrient dynamics in soil. He was elected as Fellow of National Academy of Agricultural Sciences (NAAS) and awarded ICAR-LBS Outstanding Young Scientist Award, Mosaic Foundation best Scientist in plant nutrition, Dr. K. J. Tejwani Award in the area of natural resource management

Dr. Sushmita Munda is working as Scientist at ICAR-National Rice Research Institute, Cuttack, Odisha, India. She was awarded with University Merit Scholarship, Junior Research Fellowship, and Senior Research Fellowship. She has been working on weed management under RCTs and biochar and their effects on carbon dynamics and GHGs emissions in rice. Dr. Munda published more than 30 research papers in international and national repute journals. She published a book and several book chapters, popular articles on different aspects of agriculture. She guided MSc and PhD students.

Mr. Pradeep Kumar Dash is working as a Research Associate in Crop Production Division of ICAR-National Rice Research Institute, Cuttack, Odisha, India and has 8 years of research experience. His research interests are soil organic carbon dynamics in different nutrient management practices and cropping systems, conservation agriculture, soil system functioning in relation to climate change scenario and also changes in soil microbial diversity under climate change. To his credit, he has published 2 book chapters and 20 research articles in reputed peer reviewed journals.

CLIMATE CHANGE AND GREENHOUSE GASES EMISSIONS

Pratap Bhattacharyya
ICAR-National Fellow and Principal Scientist
Division of Crop Production
ICAR-National Rice Research Institute (NRRI)
Cuttack, Odisha, India

Sushmita Munda
Scientist
ICAR-National Rice Research Institute
Cuttack, Odisha, India

Pradeep Kumar Dash
Research Associate
Crop Production Division
ICAR-National Rice Research Institute
Cuttack, Odisha, India

CRC Press is an imprint of the
Taylor & Francis Group, an **informa** business

NEW INDIA PUBLISHING AGENCY
New Delhi-110 034

First published 2021
by CRC Press
2 Park Square, Milton Park, Abingdon, Oxon, OX14 4RN

and by CRC Press
6000 Broken Sound Parkway NW, Suite 300, Boca Raton, FL 33487-2742

CRC Press is an imprint of Informa UK Limited

British Library Cataloguing-in-Publication Data
A catalogue record for this book is available from the British Library

Library of Congress Cataloging-in-Publication Data
A catalog record has been requested

ISBN: 978-1-032-00594-2 (hbk)

Preface

Firstly, we convinced ourselves that there is a need for a book on "Climate Change and Greenhouse Gases Emissions' for student at universities. A few text books on "Climate Change and Greenhouse Gases Emissions", particularly in agricultural sector are available. One of the special features of this book is that most of the data presented and discussed are the outcome of research works done by scientists both under national and international perspectives. Written in a simple language, the book presents the principles, practices along with key messages on different relevant issues on climate change and greenhouse gases (GHGs) emissions in agriculture. Chapters also contain probable questions and few solved problems. It is primarily intended for use as text book at undergraduate and postgraduate levels, but it can also be used by researchers of environmental sciences and allied disciplines. The other important feature of the book is that techniques of GHGs measurements at field level with examples are presented which could be useful for practical studies. The book should give students a good foundation in climate change studies and inspire them to take up further studies in the advanced arena of environmental sciences.

The chapters systematically cover the major areas of climate change and GHGs emissions in agriculture. The basic concept of weather, climate, climate change, climate variability, the cause and effect relationship between GHGs emission and climate change is introduced in Chapter 1. In Chapter 2, the details on causes of climate change including natural and anthropogenic, net feedback mechanisms, carbon cycle and evidences of climate change are described. The chemistry of GHGs, radiative forcing and energy balance, mechanism of major GHGs emissions from agriculture, contribution of agriculture to GHGs emissions and sector wise emissions are discussed in Chapter 3. The techniques of measurement of GHGs at field level employing manual chamber, automatic chamber, infrared gas analyser, photoacustic spectroscopy, eddy covariance techniques are elaborated in Chapter 4. It is a unique chapter which has both theoretical and practical application. This chapter also has a specified section for equations needed for GHGs estimation and few interesting solved problems using field level data. The small but important chapter is Chapter no. 5, which deals with the issues of impact of climate change on agriculture, environment and food security. This chapter also contains the predictions of climate change consequences on agriculture as a whole and impacts on Indian agriculture in particular. The heart of the book is Chapter 6, which deals with the mitigation of GHGs emissions and climate change. This chapter emphasises on the principles of mitigation of GHGs emissions, mitigation strategies, adaptation strategies to climate change, Climate Resilient Agriculture (CRA) and Climate Smart Agriculture (CSA). In the penultimate Chapter 7, we tried to critically put the economics of GHGs emissions along with the history of 'climate change policies' and the ongoing debate on climate change mitigation and food security dilemma.

In the preparation of the book, we received ungrudging help from ICAR-National Fellow Project, ICAR-NRRI, Cuttack and number of scientists and well wishers. Those whom we wish to mention are: Dr T Mohapatra, Dr S.C Datta, Dr H Pathak, Dr P Swain, Dr M J Baig, Dr, Ch. Srinivasa Rao, Dr T Adak, Dr D Bhaduri, Dr D Mandal, Dr D Chakraborty, Dr A Bhatia, Dr N Jain, Dr S Kartykeyan, Dr M C Manna, Dr R Saha, Dr D Burman, Dr S Pal Majumdar, Dr B Majumdar, Dr S Neogi, Dr K S Roy, Mr S R Padhy, Miss P P Padhi, Miss U Nandy, Mr M Das from ICAR-NARS system.

We have duly acknowledged the sources of the diagrams and tables that have been reproduced from other sources and publications.

Pratap Bhattacharyya
Sushmita Munda
Pradeep Kumar Dash

Contents

Abbreviations

AFOLU: Agriculture, Forestry and Land Use
AR: Assessment Report
ATP: Adenosine Tri-phosphate
CFC: Chlorofluorocarbon
CH_4: Methane
CO_2: Carbon Dioxide
CRA: Climate Resilient Agriculture
CSA: Climate Smart Agriculture
DAP: Di-amino Phosphate
DCD: Dicynamide
DSR: Direct Seeded Rice
ENSO: El Nino Southern Oscillation
FAO: Food and Agricultural Organization
GHG: Greenhouse Gas
GPP: Gross Primary Production
GTP: Global Temperature Change Potential
GWP: Global Warming Potential
HFC: Hydro-fluorocarbon
IPCC: Intergovernmental Panel on Climate Change
IR: Infrared Radiation
LUC: Land Use Change
MJO: Madden–Julian Oscillation
N_2O: Nitrous Oxide
NAO: North Atlantic Oscillation
NASA: National Aeronautics and Space Administration
NOAA: The National Oceanic and Atmospheric Administration
NOP: Nitro- oxypropanol
NRC: National Research Council
PDO: Pacific Decadal Oscillation
PFC: Per-fluorocarbon
RE: Ecosystem Respiration
RF: Radiative Forcing
SOC: Soil Organic Carbon
VB: Vegetation-Browning

VG: Vegetation-Greening
VOC: Volatile organic compounds
WFPS: Water filled pore space
WMO: World Meteorological Organization

1

Introduction

1.1 Weather

Firstly, what is weather? Simply, the conditions in the atmosphere above the earth is called weather. It refers to the conditions such as wind, rain and temperature, prevailing at a particular location during periods of hours or days. We often ask "what's the weather like today"? Any individual can answer to this question as it requires only a short term experience or say exposure to the atmosphere. The TV news forecasters predict weather each day and inform about the temperature, cloudiness, humidity, or possibility of storm in the next few days. Weather is localised, say, it's a hot and cloudy weather in one part of the country and dry and windy on the other part.

1.1.1 What are the factors that influence weather?

Before we try to understand the factors that influence weather, we must understand what the components that describe a weather conditions are? Is it the rains, or the air temperature or both? The answer to this question is not so easy, and it's a bit complicated. There are many components that constitute weather. Weather components include rain, cloud cover, sunshine, winds, hail, snow, sleet, storms, thunderstorms, heat waves and many more. These components are sometimes referred as weather parameters or climate elements. Each of these components define weather patterns of a particular area. These components determine the status of local atmospheric conditions. Again, as a basic definition, weather is the state of the atmosphere and majority of weather phenomena occur in the lowest layer of the atmosphere, called as the troposphere. Does it mean that ground conditions such as the topography, water bodies and vegetation of a particular area has nothing to do with it? No, it is not so. The state of the atmosphere is subjected to various processes taking place not only in the atmosphere but also in the ocean, the sea-ice, the vegetation, etc. Weather parameters or the components of weather form a sequence of events, and the influences are far reaching and they do not remain solely in the atmosphere. For example, terrestrial radiation impacts the air temperature, air pressure and humidity. The terrestrial location also influences the cloud formation. The clouds may or may not turn into precipitation depending on the outcome (cloud or rain), which is by and large governed by troposphere. Likewise, a lack of precipitation affects not only weather conditions, but also soil moisture conditions that may result in lowering of water levels (ground water table) in soil. Wind speed and direction are governed by the atmospheric pressure and air temperature. So, all these parameters or components are interrelated and the weather conditions are the outcome of their interactions. Therefore, we can say that, the three main components of weather are temperature, precipitation and solar radiation. These components are influenced by different factors.

Solar Distance

The earth's distance from the sun results in difference in temperatures up to 4°F between the poles and the equator. The tilt of the earth toward or away from the sun governs the amount of heat that part of the earth will receive from the sun. The hemisphere which is tilted towards the sun comes closer to the sun and this part experiences summer, whereas the part tilted away, experiences winter.

Latitudinal Location

The weather pattern changes with the change in the latitude. There are very little seasonal variations at the equatorial regions. The weather does not change much during the year in the equator, as the radiation received is roughly the same throughout the year. The sun is mostly direct and intense at the equator making it a hotter place. As we move towards higher latitudes (northward or southward), we receive more sunlight or less sunlight, depending on the tilt of the earth. Also, at higher latitudes, the same amount of light from the sun covers a larger area on earth's surface. As the area increases, intensity decreases, creating the extreme cold conditions at the poles.

Air Pressure

Air pressure is created when there is temperature difference between two pockets of air, or fronts. Pockets of different temperatures tend to come to an equilibrium creating air-movement and pressure. The wind formation/ movement takes place as air temperature tries to equilibrate by moving from high pressure zones to low pressure zones. The moisture laden air moves upward towards low air pressure zones and forms the clouds. The trade winds are also developed by difference in air pressure. The solar energy is most intense at the equator that causes heating up land, water as well as air. When that hot air rises, the relatively cooler air from the higher latitudes (both north and south, including polar region) try to fill up the vacuum created by the upward movement of hot air. This phenomena creates a circuit with three phases, (i) hot air-flow from the equator rises up and moves slowly towards north and south direction after splitting into two parts, (ii) towards the poles, the air gradually cools down and come down to the earth's surface and then, (iii) it flows in reverse directions and move towards the equatorial region. These are called the trade winds that maintain a constant flow of air across the earth's surface and atmosphere.

Water Vapour Presence

The water vapour is a critical component of earth-atmosphere system and it has a significant impact on weather. Water vapour, along with moisture carries a lot of heat. The movement of storm-fronts is actually the resultant of collisions between air masses having different water vapour content with varied temperatures. The presence of water vapour also affects the humidity of the region. Therefore, the places near to seas or oceans, for instance, are generally wetter and humid than the central valley or desert. Further, large water bodies also facilitates for creation of winds. The temperature gradients between land and water bodies causes the movement of breezes towards inland during the day time and out to sea/water bodies at night.

1.2 Climate

In simplest words, climate is the long-term pattern of weather in a particular area. By definition, climate is the weather conditions prevailing in an area over a long period (Shepherd 2005; Voldoire et al., 2013). The aggregated parameter of weather (range of temperature, wind, precipitation etc.) over years (at least 30 years or more) of a sizeable geographical area is termed as climate. The difference between weather and climate is a measure of time. Weather accounts for conditions of the atmosphere over a short period of time, and climate is the pattern of atmospheric condition over relatively long period of time (> 30 years). Climate even accounts for hundreds, thousands and millions of years of change.

1.2.1 Climate system

Technically, climate is defined as the average state of every day's weather condition over a period of 30 years. We also know that the weather or climate is affected by numerous processes involving in atmosphere, land, ocean etc. In the recent past, climate has been described broadly as the statistical output of the climate system. The climate system is a complex system representing the five major components, namely, atmosphere (air), biosphere (living organism), hydrosphere (water), cryosphere (ice includes permafrost) and lithosphere (upper mantle and earth's crust). A climate is originated from the climate system, using climate indices and interacting climate elements of atmosphere, hydrosphere, cryosphere, biosphere, and lithosphere. The climate system keeps evolving due to changes in its own internal dynamics as well as external events such as volcanic eruptions, and human-induced forces such as greenhouse gas emissions and land-use change. Statistical techniques/tools are used to calculate the averages for different climatic-variables. The variability is measured also to predict the frequency of occurrence of extreme events. Normally, scientists bring together data from all over the world to study the climate of whole earth through models and investigate the causes and consequences the global climate change.

1.2.2 Climate index/indices

Climate indices are referred to the estimated values used to elaborate the state and the changes in the climate systems. Very often, these are described as the statistical tools to study climatic-variations. Each climate index is developed from specific climatic-parameters for describing certain aspects of the system. Many climate indices have been developed and evaluated for different goals at various regions of the world. Mostly, they are based on empirical equations and interrelated processes that are simulated through models. The empirical equations or models use quantifiable atmospheric-parameters include air pressure, air temperature, precipitation and solar radiation and also non-atmospheric parameters (sea surface temperature or ice cover). The climate indices generally consider a climate-base period. The climate base period is actually a reference period that includes at least 30 years of data (as recommended by the WMO (World Meteorological Organization)). Presently, the most used reference climate-base period is from 1961 to 1990. The mean values of different parameters for this period are used for up scaling or future predictions.

1.3 What is climate change?

Climate change is a catch-phrase which refers to considerable and relatively persistent changes in the global climate. It is either permanent or at least a substantial shift in worldwide climate phenomena. It is mostly associated with an increase in global average temperatures. The NASA defines climate change as "A broad range of global phenomena created predominantly by burning fossil fuels, which add heat-trapping gases to earth's atmosphere. These phenomena include the increased temperature trends described by global warming, but also include changes such as rise in sea levels, melting of ice caps, shifts in flora and fauna population and their physiological changes, and extreme weather events". Climate change occurs when changes in climate system result in new weather patterns that last for at least a few decades (three or so), or may persist for thousands of years. The climate change is also broadly defined as the changes in the statistical properties (principally its mean and spread) of the climate system over long periods of time, irrespective of the causes of change (Solomon et al., 2007). Therefore, short term fluctuations (over periods shorter than a few decades) like, El Nino, does not denote as climate change. It is a part of climate variability, which will be discussed in the next section.

Process of climate change

(i) There is exchange of energies in the climate system. The climate system receives energy from the sun and from the earth's crust (in the form of terrestrial radiation), and it also gives off energy to outer space.

(ii) The earth's energy budget is governed by the balance of incoming and outgoing energies and the passages through which the energy is passed or circulated in the climate system.

(iii) If the incoming energy is greater than the outgoing energy, then the earth's energy budget is positive and warming up of climate system occurs. But if more energy goes out, the energy budget is negative and earth experiences cooling.

(iv) When the changes in the long term average are persistent over a significant period of time, it is indicated as "climate change".

1.4 What is climate variability?

Now, before we dig deep into further details of the climate change, we must understand the difference between climate variability and climate change. Climate variability is sometimes alternatively used as climate change, which is not correct. We have to go back to the basics, as we know weather and climate operate on different time scales, though the basic parameters for their measurement are almost same. Similarly, climate variability and climate change also operate on different time scales (Figure 1.1).

Short term fluctuations are recorded on seasonal or multi-seasonal time scales. On the other hand, the climate tends to change rather slowly. There are many things that can cause temperature to fluctuate around the average without causing the long-term average itself to change. This phenomenon is climate variability, and this phenomenon is usually referred to time periods ranging from months to as many as 30 years, not beyond that.

Basically, climate variability occurs when there is redistribution of energy around the globe, which leads to changes in pressure, temperature and other climate variables. The change in the quantum of energy and its distribution is the outcome of various factors. Those include the atmospheric conditions viz., amount of radiation from the sun and terrestrial conditions like, volcanic activity, terrestrial radiation etc.

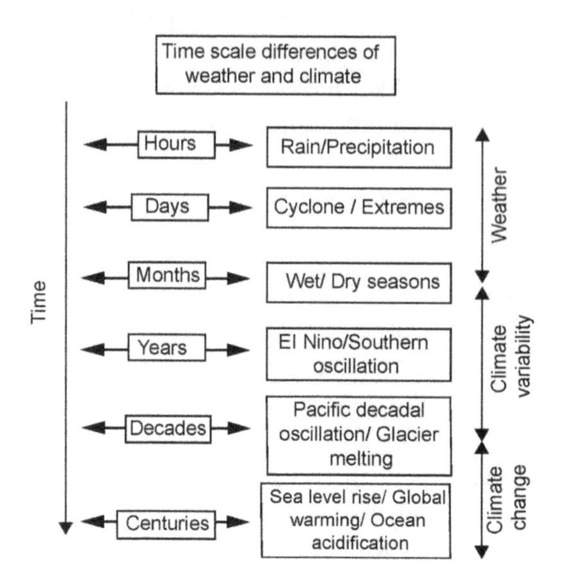

Fig. 1.1 Time scale differences of weather and climate (***Source***: https://www.pacificclimatefutures.net)

One of the most well understood causes of climate variability is the tilt of the earth, which creates seasons on earth, and the duration of the season is dependent on the latitude. One of the most prominent internal causes of climate variability is the El Nino-Southern Oscillation. It has major influence on sea-surface temperatures. The vast amount of energy is held in the oceans and even slight changes in sea surface temperatures can shift certain climate patterns. So, seasonal variations, changes in sea surface temperatures and decade long cycles of El Nino Southern Oscillation that produce warm, cool, wet, or dry periods across different regions are a natural part of climate variability. Thus, climate variability should not be confused with climate change. Only a persistent change in the average climatic conditions in long term (> 3 decades) can be considered as climate change. Most of the scientists view the changes in long term patterns in climate in recent years (after 1950) is largely man made. However, climate variability is considered very often as the outcome of natural changes in the earth's systems.

Types of Climate Variability

El Nino/La Nina-Southern Oscillation

The El Nino-Southern Oscillation (ENSO) is described as the natural variations in the ocean and atmosphere in the tropical eastern Pacific Ocean. These variations cause large-scale changes in sea-surface temperatures, sea-level pressures, precipitation and winds across much of the tropical and subtropical regions of the world. The occurrence of an El Nino condition takes place when the sea-surface temperatures of central and eastern equatorial Pacific are significantly higher than the long term average. Similarly, the La Nina conditions are just the opposite of El Nino and they occur when the sea-surface temperatures of central and eastern equatorial Pacific are significantly lower (3-5°C) than the long term average. The El Nino conditions are unfavourable for the monsoons, whereas, La Nina conditions are unfavourable for monsoons in Indian subcontinent.

Pacific Decadal Oscillation

The Pacific Decadal Oscillation (PDO) is a frequent pattern of climate variability occurring in higher latitudes (north of 20°N). The sea surface temperature anomalies (warm or cool surface waters) in the mid-Pacific basin are observed. During the warm phase, the west Pacific (northwest Pacific to be more specific) becomes cooler and eastern Pacific becomes warmer. During the cool phase, the opposite pattern is observed. During negative phase, the monsoon is affected and increased rainfall and decreased summer temperature is observed in the Indian subcontinent. The period of fluctuations in PDO is not as defined as ENSO which can continue in warm or cool phase for 10-30 years.

The North Atlantic Oscillation

The North Atlantic Oscillation (NAO) is measured by air pressure. Low pressure generally resides over Greenland/ Iceland, whereas a high pressure area is occupied in the central North Atlantic. The NAO is a weather phenomenon which refers to the fluctuations in atmospheric pressure at the sea level in North Atlantic Ocean caused by Icelandic low pressure and the North Atlantic high pressure. It controls the strength and direction of westerly winds and location of storm tracks across the North Atlantic. The areas of pressure are strengthened when NAO is in a positive phase but when the NAO is in a negative phase, the air pressure is weakened in the area. The duration of NAO may vary from days to decades, and the change in the phase is more common compared to ENSO and PDO. The positive phase is associated with more frequent and stronger winter storms across the Atlantic, and thus warm and wet winters in Europe. The negative phase is associated with more cold outbreaks in the eastern United States.

Madden-Julian Oscillation

The Madden–Julian oscillation (MJO) is one of the major components of the intra-seasonal variations in the tropical regions. It lasts for 30 to 90 days. The

MJO is a convective wave that originates in the eastern Indian Ocean and moves from west to east direction. It differs from the other forms of the climate variability oscillations, as most of them operate in a relatively smaller, defined area. The MJO consists of two phases, (i) the enhanced rainfall (or convective) phase and (ii) the suppressed rainfall phase. In the enhanced convective phase, winds at the surface converge, and air is pushed up throughout the atmosphere. When the moisture laden air rises to the atmosphere, it increases condensation and result in rainfall. In the suppressed convective phase, winds converge at the top of the atmosphere, leading the air to descend and, later, to diverge at the surface (Rui and Wang, 1990). As air dries up when it descends from high altitudes, resulting in reduced rainfall. Strong MJO activity divides the earth into two halves or dipole (two opposing centres of action); one half within the enhanced convective phase and the other half in the suppressed convective phase. These two phases produce opposite changes in clouds and rainfall and this entire dipole propagates eastward. So, torrential rainfall and dry weather periods are brought by MJO when it moves from west to east direction, returning to its point of origin after 30-60 days.

1.5 Is the climate really changing?

The atmosphere is highly complex and non-linear in how it reacts. All the elements or components of the global climate are connected. Modern catastrophe models, deals with both statistical and meteorological approaches that consider both climate change and climate variability. Climate change and climate variability operate almost in unison and each mechanism plays a role in shaping our climate. Models are regularly updated and incorporate clearly established findings using a more recent period of data to reflect risk in the current climate regime. Any climate change that has already occurred is indirectly accounted for and future changes will be taken into account as the models are updated. These models help in working out how the changes caused by fluctuations in the climatic-components that affect the entire system and appear in different areas.

The most compelling evidence of climate change is the long term data related to atmospheric CO_2 levels and global temperature. Fossil records are also being used to contemplate the change in climate. The data on atmospheric CO_2 levels, the fossil records, along with the data on shifts in distribution of flora and fauna show a strong correlation between CO_2 levels and temperature.

The other evidences include the rapid retreat of glaciers and melting of the polar ice cap. Sea levels are rising significantly and the ocean temperatures are also rising substantially. The movement of a number of fish species due to change in ocean temperature is also another indirect indication of undisputable change in climate.

Despite these evidences, some of the scientific community as well as organizations remain "climate sceptics" and doubt the theory of human made global climate change. Most of the climate sceptics believe that the apparent changes in the climate are part of climate variability and its extrapolation through the models is inconsistent and doubtful. The sceptics believe that the current models are too basic to capture such a complex process. They also believe that the available data is insufficient to support such big claims. A group of scientists think that greenhouse data is insufficient and not applicable in earth's complex systems for CO_2 is too insignificant to change the overall climate of the world.

Following are some of the evidences/ findings of climate change as reported by IPCC

1. There has been change in the CO_2 concentration in the atmosphere, partly due to natural factors and largely due to human induced activities, for example, exhaustive use of fossil fuels. The rise in temperature has been about 0.85°C since the late 19th century, which is considerably a very short period bearing in mind the formation and evolution of earth over ages. Past few years have been the warmest years in the climate's recorded history, with the five warmest years on record taking place since 2010. Not only the annual temperatures but also the monthly temperatures have been showing an increasing trend compared to the long term averages.
2. The oceans temperatures are also increasing along with the land-surface temperatures, particularly at the top 700 meters (2,300 feet) of ocean showing warming of more than 0.33°C since 1969 (Roemmich et al., 2012; NOAA, 2019).
3. An average of 286 billion tons of ice is lost per year between 1993 and 2016 in Greenland, while Antarctica alone lost about 127 billion tons of ice per year during the same time period (NASA's, Gravity Recovery and Climate Experiment). The rate of Antarctica ice mass loss has tripled in the last decade.
4. Retreating of glaciers is the most authentic record of climate change. It has occurred worldwide, in all the mountain ranges including in the Himalayas, Alps, Rockies, Andes, Africa and Alaska.
5. Satellite images reveal that the Northern Hemisphere is witnessing an accelerated rate of decline in the amount of spring snow cover in the past five decades.
6. The rate of increase in the sea level in the last two decades is almost double to that of the last century and more alarming is that the rate is increasing every year.

7. Reports have also suggested a rapid decline in Arctic sea ice in the last few decades.

1.6 The cause and effect relationship between GHGs emission and climate change

Greenhouse gases can absorb and radiate back the range of infrared (IR) radiation and thereby effectively absorb as well as emit heat energy and keeping our earth-atmosphere warm. The naturally occurring GHGs include CO_2, CH_4, water vapour, and nitrous oxides, while others are synthetic. Those that are man-made include the chlorofluorocarbons (CFCs), hydro-fluorocarbons (HFCs) and per-fluorocarbons (PFCs), as well as sulfur hexafluoride (SF6).

The "greenhouse effect" is the process by which greenhouse gases in atmosphere absorbs and radiated back the long-wave radiations (preferably IR radiations) in earth-atmosphere and warms our planet surface temperature. The surface temperature would be around 256°K or -17°C without any greenhouse effect and there would be no existence of life in our plant without that process. However, due to the sustained greenhouse effect in our planet the present day mean global surface temperature is around 14.85°C (IPCC, 2014). But unfortunately, the mean global surface temperature has been increased by 0.85°C in last century due to "enhanced greenhouse effect". 'Enhanced greenhouse effect' causes global warming which has several consequences (both positive and negative) on agriculture, environment, industry and other sectors of our society.

The 'enhanced greenhouse effect' on earth-atmosphere is the resultant of increase in GHGs emissions due to anthropogenic activities. The 'enhanced greenhouse effect' is either caused by increased well known GHGs (water vapour, CO_2, CH_4, N_2O) present in atmosphere or by addition of some other radiative gases which is not naturally presents (e.g. HCFC-hydro-chlorofluro carbon, CFC, etc). The global atmospheric temperature is increasing due to this 'enhanced greenhouse effect' which is called global warming. Environmentalists and climate scientists think that the human induced greenhouse effect is the prime cause of current trend of global warming. This thought is further strengthened as the present-day climate models could not explain the observed trends of temperature rise with the help of solar irradiation only and more so warming is only experienced at lower atmosphere.

Moreover, the rises in the human induced greenhouse gases over the last few centuries have been alarming. This has been caused by the industrial and technological revolution owing to an overall increase in global population, stimulating more greenhouse gases causing global warming.

Climate change, on the other hand, refers to a broad range of global phenomena in which global warming plays an important role. It also acts as an indicator to climate change. Climate change phenomena include the increased temperature trends indicated by global warming, but also include changes such as rise in sea level due to melting of ice caps and glaciers worldwide; shift in agricultural systems; and various extreme weather events.

Put simply, while the increase in atmospheric temperature over the long term averages is more precisely referred to as global warming, the climate change not only includes earth's increasing global average temperature, but it also includes the other long term changes such as increase in frequency of occurrence of extreme events due to increase in global average temperature. Climate change is the term currently favoured by researchers, as it encompasses all the irreversible or persistent changes in the climate.

The GHGs emission is the cause of global warming leading to climate change. However, climate changes also sometimes promote higher GHGs emission. In nutshell, we can say, the global warming is the one of the major drives climate change.

1.7 What are the causes of climate change?

There are both natural as well as anthropogenic causes of climate change. It could be the resultant of internal variability in the climate system, when natural processes such as cyclical ocean patterns (El Nino–Southern Oscillation, Pacific decadal oscillation, etc.) persist for a significantly longer period and affects the earth's energy budget. Climate change can also results from external energy sources, like changes in solar output and alteration of volcanic produce within the system itself.

Many environmental and climate scientists also believe that anthropogenic activities are also a significant cause of climate change in earth's-atmosphere system. Human activities plays an important role for driving climate change through global warming in last century. There is a wide range of natural phenomena that can affect the climate. But most of the climate scientists believe that the recent global warming and resultant climate effects that we're witnessing are the resultant of human activity. The details of causes of climate change, GHGs emissions, greenhouse effect and global warming and their interrelationship would be discussed in coming chapters of this book.

1.8 Key messages

1. The weather and climate are different term, primarily in respect of time scale.
2. Climate change and climate variability are also different term on time scale basis.
3. Weather and climate variability are relatively short term observations of meteorological variables. While, climate and climate change are long-term phenomena.
4. There are natural as well as anthropogenic causes of climate change.
5. Greenhouse gases (GHGs) can absorb and radiate back the range of infrared (IR) radiation and thereby absorb as well as emit heat energy and keeping our earth-atmosphere warm.
6. The "greenhouse effect" is the process by which GHGs in atmosphere absorbs and radiated back the long-wave radiations in earth-atmosphere and warms our planet.
7. There would be no existence of life in our plant without the process of "greenhouse effect".
8. The GHGs emissions and global warming are the drivers of climate change. However, climate change also affects the GHGs emission in different systems.

1.9 Probable questions

1. Differentiate between weather and climate.
2. How the climate variability differ from climate change?
3. Write down the causes of climate variability.
4. What do you understand by climate system?
5. What are the causes of climate change?
6. Define GHGs, greenhouse effects and global warming.
7. Write the primary cause-effect relationship between GHGs emissions and climate change.
8. Do you think that climate is changing? Support your answer.

References

Bindoff NL, Willebrand J, Artale V, Cazenave A, Gregory JM, Gulev S, Hanawa K, Le Quéré C, Levitus S, Nojiri Y, Shum CK. Observations: oceanic climate change and sea level.

https://www.pacificclimatefutures.net/en/help/climate-projections

IPCC (2014), Climate change 2014: Impacts, adaptation, and vulnerability. Contribution of Working Group II to the Fifth Assessment Report of the Intergovernmental Panel on Climate Change, edited by C. B. Field, V. R. Barros, D. J. Dokken, K. J. Mach, M. D. Mastrandrea, T. E. Bilir, M. Chatterjee, K. L. Ebi, Y. O. Estrada, R. C. Genova, B. Girma, E. S. Kissel, A. N. Levy, S. MacCracken, P. R. Mastrandrea, and L. L. White, Cambridge Univ. Press, Cambridge, U. K.

NOAA National Centers for Environmental information, Climate at a Glance: Global Time Series, published June 2019, retrieved on July 3, 2019 from https://www.ncdc.noaa.gov/cag/.

Roemmich D, Gould WJ, Gilson J. 135 years of global ocean warming between the Challenger expedition and the Argo Programme. Nature Climate Change. 2012 Jun;2(6):425.

Rui H, Wang B. Development characteristics and dynamic structure of tropical intraseasonal convection anomalies. Journal of the Atmospheric Sciences. 1990 Feb;47(3):357–79.

Shepherd, J.M., 2005. A review of current investigations of urban-induced rainfall and recommendations for the future. *Earth Interactions*, *9*(12), pp.1–27.

Solomon S, Qin D, Manning M, Averyt K, Marquis M, editors. Climate change 2007-the physical science basis: Working group I contribution to the fourth assessment report of the IPCC. Cambridge university press; 2007 Sep 10.

Voldoire A, Sanchez-Gomez E, y Mélia DS, Decharme B, Cassou C, Sénési S, Valcke S, Beau I, Alias A, Chevallier M, Déqué M. The CNRM-CM5. 1 global climate model: description and basic evaluation. Climate Dynamics. 2013 May 1;40(9–10):2091–121.

2

Causes of Climate Change

Climate change is caused by both natural as well as anthropogenic causes and it has been changing since the inception of earth-atmosphere (Figure 2.1). In a broader sense, anthropogenic activities (human activities) also must be considered as natural one in longer time scale as human and their activities are also a part of natural-ecosystem-functions. Among the natural phenomena, volcanic eruptions and continental drifts significantly affect the earth's precipitation and temperature. A huge volume of CO_2 is released due to volcanic eruption at a little span of time-period which heated up the earth-atmosphere. Climate variability is primarily governed by rise in sea temperature, movement of warm water from the Western Pacific towards Peru, and El Nino which occurs at an intervals of 3-7 years. Climate variability also influenced the alteration in temperature and precipitation world wide. And persistent climate variability cause long term climate change. Therefore, basically three drivers that causes climate change are, the alteration in the energy of sun received by earth-atmosphere (natural causes), variations of reflectivity of earth-surface-atmosphere system (natural as well as anthropogenic causes) and fluctuations in the greenhouse effect (natural as well as anthropogenic causes).

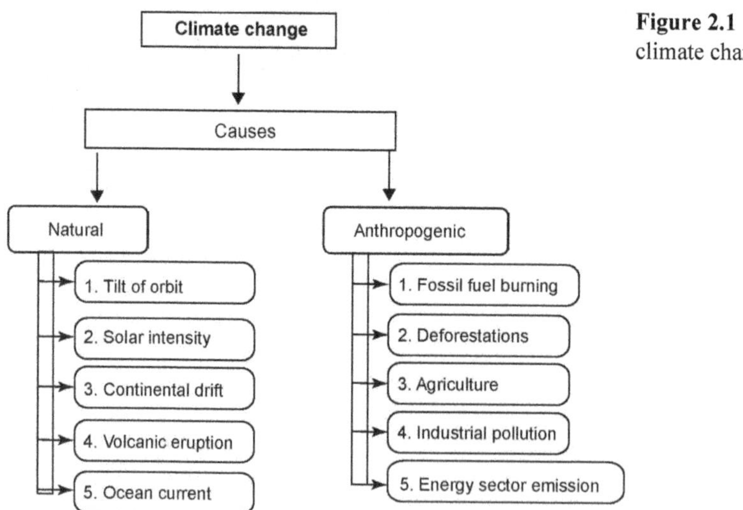

Figure 2.1 Causes of climate change

2.1 Natural causes of climate change

Alteration in solar energy

One of the primary causes of natural climate change is alteration in solar energy intensity received by earth-atmosphere. The fusion and fission leading to emission of high intensities of energies is the inherent feature of sun. The alteration of sun's own structural get up has been changing with time which leads to fluctuations of emitted energy outputs by sun. As we know, the cooling generally occurs on earth during the period of weaker solar intensity and warming of earth at the time of stronger solar intensity. Usually there is a natural 11-year cycle of ups and downs in solar intensity.

Tilting of earth orbit

The tilting of earth orbit, its position and shape, earth axis orientation also directly influence the quantum of radiation (sunlight/ energy) received by earth-surface. Those are important drivers of natural climate change. There was cooling effect due to reduced solar energy-intensity during 17-19th century (little ice age) making the world climate cooler, specifically in North America and Europe. After that, there was shorter warming period in intermediate ice age. Those were basically happened due to tilting of earth orbit as well as changes in solar intensity. The long term effect of these phenomena on climate change should be judged at larger time scale.

Volcanic eruptions

In volcanic eruptions, the hot substances from the earth's core are thrown out to the earth-surface. In this process a large amount of sulphur dioxide is also thrown out to the lower stratosphere that forms aerosol particles that produce dust veils which last several years and cool the earth's surface. At the same time, those dust veils warm the stratosphere by absorbing higher amount of solar radiation. Therefore, volcanic eruption influences the greenhouse effect both of the troposphere and the stratosphere. Actually, the net radiative effect of the volcanic aerosols is to cool our planet. But, the precipitation fluctuations due to volcanic eruptions are likely to have more direct impacts on society and the environment.

Continental drift

The episodic convergence of major continental plates is responsible for formation of giant super-continents, like, Rodinia, Gondwana, Pangea, and Laurasia in past. In the past, those enhanced continents have produced extreme seasonal temperature changes and aridity in the interior parts of continent, while typical monsoonal-circulation patterns in several coastal regions. As for example, in central Pangea, mean summer temperatures were around 6-10°C warmer than that of today's temperature (Crowley and North, 1996). Aridity prevailed at the continental-interiors as majority of the atmospheric water vapor was lost to rainfall near the continental-margins. Further, enhanced seasonality and a less precipitation in the continental-interior of ancient supercontinents were unfavorable for the growth of continental-scale ice sheets in mid-high latitudes. The summer cooling of Antarctica in the Paleogene was thought to be caused by the Cenozoic separation of Southern Hemisphere continents that also associated with the declining of CO_2 concentration, reductions in seasonality and continentality.

2.2 Anthropogenic causes of climate change

The anthropogenic activities basically influence the 'greenhouse effect' and change the reflectivity of earth-surface-atmosphere system which consequently causes climate change. Evidences showed that climate changes before the 'Industrial Revolution (1750)' was primarily due to natural causes. However, changes occurred in climate during middle of 20th century to recent years mostly due to anthropogenic activities. We all have been knowing that anthropogenic activities (primarily human activities) are affecting the climate since the 'Industrial Revolution'. As for example the average global surface temperature has been risen by ~1°C (IPCC, 2014) since industrial revolution. The average rise of sea level is around 20 centimeter. If we are more specific about time period, then one can say the influence of nature on climate change was more significant than human activities up to 1950. The trend of climate change after that only could be reasonably explained through sector wise activities like, power sector, energy sector, vehicular pollution, fossil fuel burning, land use change, agricultural activities, etc. Those are indicative of greater influence of anthropogenic activities over natural phenomena as the causative agent of climate change. Further, the 5th assessment (AR5) report of IPCC (2014) clearly stated that more than 90% evidences showed climatic changes and global warming in last century are caused by increased GHGs in atmosphere emitted due to anthropogenic activities. The sector wise anthropogenic activities vary and more so the GHGs emission. The magnitude of emissions from different sectors also varies and keep changing with advancement of technologies and mitigation approaches followed. It also varies with national and international policy changes on emission rules. Therefore, the extent of global warming in future is uncertain and more so the quantum of climate change. Further, we do not know, how much GHGs would be going to increase in future which depend on socio-economic growth and technological advancement. Above all, the resilience of our climate system and its sensitivity and buffering capacity are still uncertain and poorly understood by scientific communities.

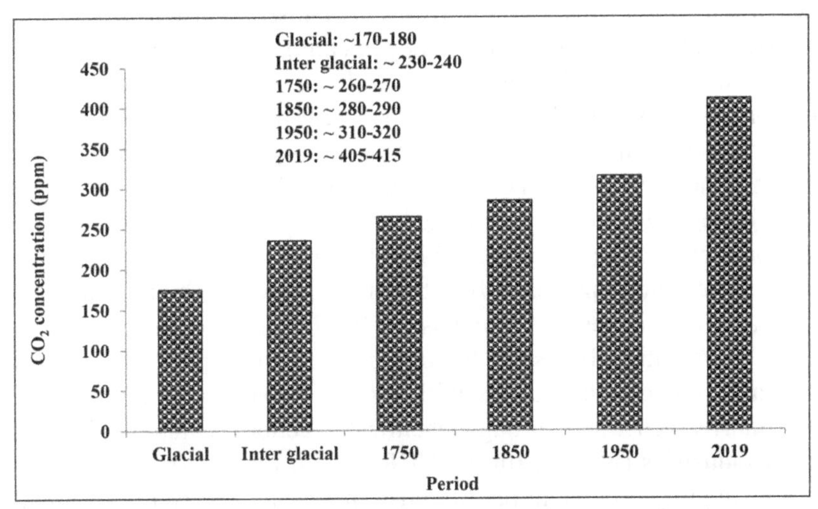

Figure 2.2 The carbon dioxide concentration in earth-atmosphere since ice age.

Changes in greenhouse effect

Three major factors *viz.*, (i) the glacial cycle, (ii) heating-cooling of ocean and (iii) human activities significantly influence the 'greenhouse effect'. The CO_2 and H_2O vapour which are two major greenhouse gases are affected significantly by above mentioned three factors. In the previous millions of years the earth-surface CO_2 levels in the planet basically varied with glacial cycles. The CO_2 concentration in atmosphere was lower in "cool glacial" age then it was increased in "interglacial" period (Fig 2.2). The heating-cooling cycles of land surface and ocean also altered the GHGs concentration in earth-atmosphere and influenced the "greenhouse effect'. However, the so called "enhanced greenhouse effect" as seen by last 200-250 years since the 'Industrial Revolution (1750)' was primarily caused by human activities. Those 'enhanced greenhouse effect' not only caused global warming but also changed our climate a lot. For instance, the atmospheric CO_2 concentrations have increased around 40% since pre-industrial times. It increased from 280 ppmv in the 18th century to around 412 ppmv in 2019 (NOAA-ESRL, 2019). In matter of fact, the present CO_2 concentration in earth-atmosphere is more than that, what it has been in past 80 thousand years.

The anthropogenic activities causing the 'enhanced greenhouse effect' could be listed as; (i) power sector emission, (ii) fossil fuel burning, (iii) vehicular pollution, (iv) industrial emission, (v) land use change, (vi) deforestation and (vii) unscientific agricultural activities (Figure 2.1). Human activities presently responsible for 135 times more CO_2 emission (30 billion tones yr^{1}) than that of volcanic eruptions per year (NRC, 2010; USGCR, 2014). Methane (a GHG), which is naturally produced from wetlands and fossil fuel extraction and also from rice-paddy cultivation (anthropogenically) was also measured (in 2019) highest since past 80 thousand years. In last century (20th century), its concentration has grown by ~2.5 times over pre-industrial period. Fortunately, the rate of increment of methane concentration in atmosphere has reduced in last 3 decades. Another important GHG is nitrous oxide (N_2O) which is naturally produced in biological systems. It is also generated through indiscriminate use of nitrogenous fertilizers (like urea, DAP, GROMOR, ammonium sulphate, ammonium nitrate etc.) and fossil fuel burning. Therefore, human-induced processes play a major role in N_2O emission. Its concentration has also been increased by ~20% since 'Industrial Revolution'. The tropospheric ozone and volatile organic compounds (GHGs with higher global warming potential (GWP)) primarily generated through chemical reaction in automobile and power-plant industries (NRC, 2010; USGCRI, 2014; IPCC, 2014). However, water vapour the most abundant natural GHG has generated and maintained in earth-atmosphere through natural processes.

Variations of reflectivity in earth-atmospheric system

The terrestrial radiation is generated by reflection from earth-components. We know, some portions of incoming radiation from sun absorb by earth surface and cloud and rest reflected back to the system (earth-atmosphere). The earth-components include soils, stones, deserts, ice-cover, forest, agriculture-crops, ocean surface, other vegetation cover, etc. These components vary with their surface properties like surface roughness, colour, aspects and chemical makeup. These properties directly and indirectly govern the adsorption and reflection capacities of earth-atmosphere as absorption-reflection is a surface phenomenon. As for example, the light coloured white snow and smooth surface like clouds reflect higher quantum of light. On the other hand, dark coloured brown/black-soil, ocean and forest absorb majority of incident sunlight. The ratio of reflected to incident radiation is called albedo. It is used to measure reflection capacity of earth-components. The albedo of white snow cover at pole region is usually > 85%, whereas, black soil or forest has albedo < 15. Overall, the earth albedo is around 30%.

Aerosol, cloud cover (also type of clouds), haze also plays an important role in earths' albedo, hence regulating the reflectivity of earth-atmosphere and impacted climate change. Aerosol formed by particles erupted from volcano and sulfur emitted from coal burning reflected back a considerable portion of incoming solar radiation, hence have a cooling effect on earth-atmosphere. So, we see part of the cause of this type of aerosol formation is natural and rest is anthropogenic. Therefore, fluctuation of reflectivity of earth surface which cause climate change, partly influenced by natural phenomena and partly by human activities. Natural causes of change in reflectivity of surface-component which have both warming and cooling effect in the past was very slow process and primarily driven by melting of sea-ice and aerosol formed by volcanic-eruption. Present day rapid change in climate is governed by quick alteration in reflectivity of surface object caused by the changes of land covers, land uses, desertification, deforestation urbanization, and human-induced generation of aerosol. Human induced production of aerosols also cool the earth-atmosphere and sometimes counter act the warming processes. Therefore, now-a-days to explain the causes of climate change the net feedback mechanism is getting more importance than the single factor-explanation.

2.3 Net feedback mechanisms: The cause of climate change

The feedback which facilitates the climate change is called positive feedback, while, if hinders the changes is termed as negative feedback. Majority of environmentalists think that the net feedback mechanism of different drivers of climate change is more valid than a single driver. Net feedback could enhance or hinder the cooling or warming effect caused by single driver of climate change. For an instance, the water vapor, carbon dioxide, methane and nitrous oxide are the important GHGs that absorb infrared radiation which radiated from the earth-surface to warm our atmosphere. The surface then returns back portion of this heat to the air. The carbon cycle-thermostat is a negative feedback mechanism to climate change that is based on chemical weathering, carbonate mineral formation and volcanism. The chemical weathering and carbonate formation increase with warming of earth-surface which is consequently drawing-out CO_2 from of the atmosphere and cooling the planet by reducing the greenhouse effect. On the contrary if the earth-atmosphere cools, the process of carbonate formation and chemical weathering is slowed down and subsequently the withdrawal of CO_2 from the atmosphere becomes less. So, here the net feedback of these two contrasting mechanism would decide the direction and intensity of global warming and climate change (Fig 2.3). Further, the volcanic eruptions could add CO_2 to the atmosphere and warm our planet by increasing the greenhouse effect. But at the same time, if the particles erupted from volcano forms aerosol (radiated back the incoming solar radiation to space) then it cools the atmosphere. Here also the net balance of two processes would provide a net feedback that actually governs the extent of climate change. Similarly, the net feedback of different factors like plate tectonic movements of continents, solar luminosity, the earth's orbital tilt, the rate of sea floor spreading, ocean circulation, volcanism, mountain building, etc., that influence the climate on the earth need to be considered simultaneously. As for example, in ancient times the earth was warm in spite of a weaker sun due to high GHGs concentration in atmosphere. Afterwards, the energy output form sun gradually increased over time and the carbon cycle thermostat maintained a temperature where organisms could survive. The process primarily happened by storing of carbon as carbonate rock in the earth crust by removing carbon dioxide from the atmosphere.

Another important example of feedback mechanism is that, although evaporation has a cooling effect but water vapour (being a GHG), enhance global warming and provide a positive feedback. On the other hand, some type of high reflective cloud causes negative feedback to warming (reflected more sunlight to space). Therefore, permafrost thawing, ocean absorption of CO_2, sulphur-particle produced aerosol, relative GHGs concentration as well as biotic and micro-organisms diversification in soil must be considered when judging the processes for positive and negative feedback to climate change. Therefore, feedback mechanism is taken into consideration of all significant causes/ factors to climate change like solar intensity, reflectivity, aerosols behaviour, GHGs concentration, carbon cycle etc., at a particular time.

2.4 The carbon cycle

Carbon is the fourth most essential element in the universe and ranked number 6 in the periodic table. Mainly, it is the basic building block of all organic molecules. Carbon is omnipresent in all the biological molecules including dead and alive and basically combines with oxygen, hydrogen, nitrogen, phosphorus, sulphur etc. It is an integral part of cellulose, sugars, fats, nucleic acids, and adenosine triphosphate (ATP), an energy transfer molecule. It acts as a sink for several forms of carbonaceous compounds in earth-atmosphere systems. The basic building blocks of both living and dead materials are made up of carbon. In the atmosphere, it is mostly found as GHGs molecules like, carbon dioxide and methane. However, these two gases are less than 0.1% by volume of the earth-atmosphere system but contribute significantly in 'greenhouse effect'. In the hydrosphere, bicarbonate (HCO_3^{-1}) ion is the most common form of carbon. This ion is formed by the dissolution of carbon dioxide in water, primarily in the ocean. Carbon as CO_2 gas, dissolved in ocean-water is about 50 times more than that of atmosphere. Apart from atmosphere and hydrosphere, carbon is also part of carbonate minerals making up the rocks like limestone, dolomite and marble. It is also present in the earth-crust as hydrocarbons derived from fossil fuels (like coal, oil and natural gas). Apart from this, carbon is very much present in the form of dissolved CO_2 gas in the earth's mantle and in magmas.

The carbon cycle is the combination of different bio-chemical pathways through which carbon flows between the earth systems. Broadly, carbon moves through earth's atmosphere, biosphere, geosphere, and oceans to maintain ecological balance and climate of our planet (Figure 2.4). Carbon cycle is the combination of complex processes. Therefore, it is convenient to study the sink-source relationship and pathways involving organic carbon separately from that of inorganic carbon. The organic carbon cycle is relatively shorter time-scale that includes the biologically mediated processes and operated by human and animals, while, the inorganic carbon cycle includes non-biological processes. So, the inorganic carbon cycle operates over longer, geologic spans of time.

Photosynthesis is the most important biochemical process that exists in earth-atmosphere. In which, the plants use the solar energy (light energy) and produce carbohydrate with the help of water and carbon dioxide. In the process of photosynthesis the oxygen is liberated in atmosphere by not only plants but also by algae and cyanobacteria having chloroplast. In photosynthesis, light energy is converted to chemical energy. Chlorophyll molecules play an important role for capturing the light energy and then use a trans-membrane pH gradient to synthesis the ATP in the living cell. Photosynthesis provides the energy to reduce carbon required for the survival of virtually all living systems on our planet. Photosynthesis not only help in production of food but at the same time creates molecular oxygen, which is the basic need of living organism. The respiration process is opposite to photosynthesis. In respiration, organic molecules are broken down and chemical energy is produced and by this process carbon dioxide is released to the atmosphere. These two processes are in equilibrium in living systems. Each molecule of CO_2 taken out from the atmosphere by photosynthesis is returned back to atmosphere by the process of respiration during the metabolism of organic carbon. Similarly, each oxygen molecule released into the atmosphere by the process of photosynthesis is taken up during respiration.

The biological substances must die and subjected to decay. The decay of organism could be taken place in aerobic or anaerobic condition or may be a cycle of aerobic and anaerobic conditions and vice versa. The scavengers, bacteria and fungi are primarily responsible of aerobic catabolism, while, anaerobic catabolism is mostly done by bacteria and archaea. The carbon dioxide or methane is produced from organic substances after decay and releases chemical energy and heat. Subsequently, CH_4 chemically reacts with O_2 in the atmosphere and produces CO_2 and water. So, the cycle is on. The end result of both respiration and decay is same, i.e., CO_2.

The earth-atmosphere's oxygen is 23%. In world, oxygen mass is half of the total standing biomass of organic carbon. However, the production of oxygen by the process of photosynthesis is around 2.5 times than that of biomass carbon. Therefore, the main cause of the excess of O_2 in the earth-atmosphere is that a considerable amount of organic carbon produced by photosynthesis is stored for a longer time scale in a variety of sinks. And only a part of that is recombining with oxygen to produce carbon dioxide. Now the question is where is the missing organic carbon? The scientific answer of that majority of that organic carbon is stored in bogs, swamps, soils, and lakes on land. There is 3 to 5 times more organic carbon existing in soils compared to vegetation and fauna. Similarly, in the oceans, carbon produced by marine organisms by photosynthesis is absorbed into the oxygen-poor water at the deep sea. It is then dissolved in cold sea water in deep and stored for a long time in marine mud/ sediment. This mechanism through which carbon moves from biologically active sea-surface water to the deep water and stored for long time is called the biological pump (Figure 2.4).

During the formation of sedimentary rock on earth-crust, the organic carbon both in soil and ocean are trapped. This accumulating sediment can store the carbon for thousand to hundred millions of years. This is one of the important ways by which the sequestration of organic carbon in the earth crust is taken place. Similarly, in swamps, bogs and peat on land surface the plant biomass accumulating a large amount of organic carbon and form a thick layer of organic matter on surface. When this peat/bog layers having high concentration of organic carbon get buried under relatively younger layers of sediments, coal is formed after several years based on pressure exerted and temperature that prevailed in the specific region. Organic carbon also accumulates by algae and planktons and deposited on the sediments / mud present in lakes and oceans. If these sediments / mud become buried, then the trapped carbon molecules would break down under intense pressure and heat and form hydrocarbons like, natural gas, tar, oil, etc. In fossil fuel, actually, sedimentary organic carbon produced by ancient photosynthesis having high concentrations of energy collected from solar radiation ultimately converted to chemical energy and stored for longer time. So, fossil fuels contain carbon which is stored over hundreds of millions of years of geologic time. Therefore, if we burn fossil fuels in power plants, vehicles, and homes to have energy, it actually releases carbon which was stored for millions of years in earth, and allows it to move into the atmosphere as CO_2. Unfortunately, in last hundred years or so, we have extracted and burned huge quantities of fossil fuels that have accumulated over millions of years. In this process, we are directly sending several billions of tons of ancient carbon stock back into the earth-atmosphere every year. We must remember, that the fossil fuels are non-renewable resources of energy as once they are exhausted, they could not be replenished on a human-time scale. More so, we can not wait for millions of years for the geologic processes that bury carbon in the crust of the earth to produce new hydrocarbon deposits. So, we must be careful and try to use renewable sources of energy like solar radiation, wind, water flow, crop residues to protect our mother earth for long time.

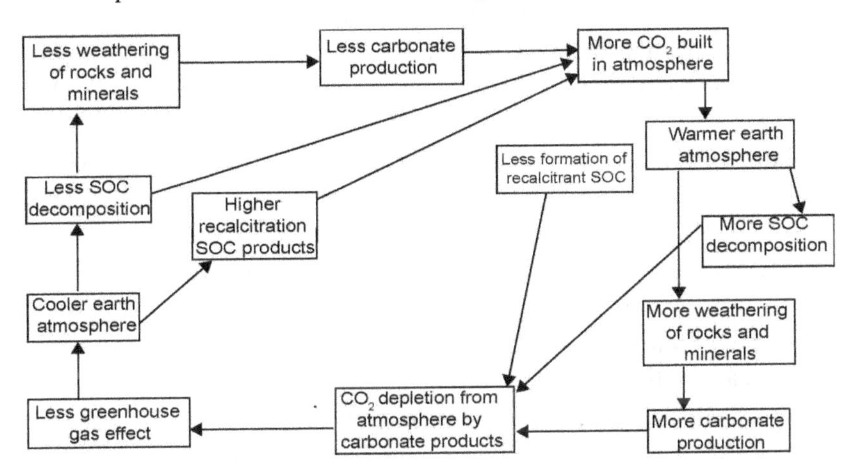

Figure 2.3 The feedback mechanism of carbon cycle-climate change.

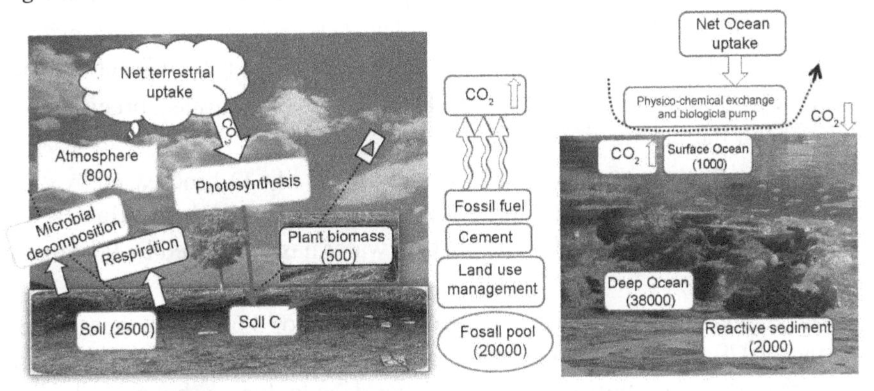

Figure 2.4 The carbon cycle in earth-atmosphere
(*Note: The values in parenthesis in the figure is Gt-C*).

2.5 Evidences of climate change

Many literatures and workshop/ symposium enlisted the evidences of climate change on the earth system. In this section we have listed the most cited evidences given by most authentic agency in the world (i.e., Intergovernmental Pannel on Climate Change (IPCC)) studying the climate change issues. We have considered the 5th Assessment report of IPCC (2014) and recent interim report published on 2018.

The evidences of climate change

1. The global average surface (combined of land and ocean) temperature has been increased by 0.85°C (ranged from 0.65 to 1.06°C) over the period 1880 to 2012 (IPCC, 2014).
2. The latest data from IPCC (2018) stated that the long-term warming trend of global mean surface temperature since pre-industrial times is for the decade 2006-2015 was 0.87°C higher than the average over the 1850-1900 periods.
3. The last three decades also has been consecutively warmer on earth-surface (combined (land and ocean)) than those of preceding decade since 1850.
4. In the Northern Hemisphere, the warmest 30-year period of the last ~1400 years was the time-period from 1983 to 2012.
5. The rate of warming during 1998 to 2012 (15 years period) was 0.05°C per decade with a range of -0.05 to 0.15°C.
6. The warming of ocean was highest near the upper 75 m surface by 0.11°C per decade over the period 1971 to 2010 on a global scale.
7. Warming was generally higher over land surface than over the ocean.
8. Human-induced warming reached ~1°C (±0.2°C) above pre-industrial levels in 2017, and was increasing at the rate of 0.2°C (±0.1°C) per decade.
9. In the mid-latitude of the Northern Hemisphere (land areas), precipitation has been increasing since 1901.
10. The acidification of the ocean (through uptake of CO_2) increased by 26% as measured by hydrogen ion concentration since the beginning of the industrial era. The pH of ocean surface water has decreased by 0.1 during the same time.
11. The Arctic sea-ice extent decreased at the rate of 3.5 to 4.1% per decade since 1979 to 2012. However, the annual mean Antarctic sea-ice extant increased in the range of 1.2 to 1.8% per decade during that period.
12. Glaciers shrink almost worldwide. Northern Hemisphere spring snow cover has also been decreasing in extant. At the same time the permafrost temperatures have increased in most regions since the early 1980s in response to changing snow cover and increased surface temperature.
13. The global mean sea level rose by 0.19 m during the period 1901 to 2010. The rate of rise since the mid-19th century has been higher than the mean rate during the previous two millennia.

14. Trends in occurrence of very intense tropical cyclones (category 4 / 5 hurricanes on the Saffir-Simpson scale) increased over recent decades.
15. An increase in the frequency of extremely severe cyclonic storms reported, however, tropical cyclones and severe tropical cyclones exhibited decreasing trends (over the period 1961-2010).
16. Global warming is intensifying hurricanes and other tropical weather systems and is also changing the path of those storms.
17. In 2017, the heating effect due to greenhouse gases (measured through "greenhouse gas index" by NOAA) increased by 1.6 percent and since 1990, the human-caused emissions "turned up the warming" by 41 percent in last three decades.
18. The global populations of vertebrate species have declined by 60 percent (in size) in the past 40 years.
19. Hundreds of fish species have been moving towards the north to cooler water in the oceans and disrupting coastal economies.

2.6 Key messages

1. Climate change is caused by both natural and anthropogenic factors.
2. The anthropogenic factors were more dominant after 1950's as evident from IPCC data inventories.
3. The three drivers that cause climate change are the alteration in the energy of sun received by earth-atmosphere, variations of reflectivity of earth-surface-atmosphere system and fluctuations in the greenhouse effect.
4. The anthropogenic factors causing climate change include fossil fuel burning, energy and power sector emissions, land use changes, faulty agricultural practices, vehicular pollution and deforestation.
5. The natural factors causing climate change include changes in solar radiation intensity, earth-orbital tilt, volcanic eruption, continental drift and ocean-current fluctuations.
6. Net feedbacks of different drivers to climate change actually govern the intensity and direction of climate change.
7. Carbon cycle-climate change feedback is the key to maintain ecological balance on earth-surface.
8. Atmospheric and oceanic temperature increase, sea level rise, melting of glaciers, increase of extreme weather events and ocean acidification are the most cited evidences of climate change.

2.7 Probable questions

1. What are the drivers of climate change?
2. Differentiate the natural and anthropogenic causes of climate change.
3. How the greenhouse gases emissions regulate the climate change?
4. What is the Carbon cycle in earth-atmosphere?
5. Explain why the net feedback of different drivers to climate change is more important to judge the direction and intensity of climate change.
6. Establish the relationship of terrestrial carbon cycle and climate change.
7. Draw a neat sketch of the carbon cycle on earth-atmosphere.
8. Write down the 10 most cited evidences of climate change as stated by IPCC.

References

Intergovernmental Panel on Climate Change, 2018. Intergovernmental Panel on Climate Change. Global Warming of 1.5° C: An IPCC Special Report on the Impacts of Global Warming of 1.5° C Above Pre-industrial Levels and Related Global Greenhouse Gas Emission Pathways, in the Context of Strengthening the Global Response to the Threat of Climate Change, Sustainable Development, and Efforts to Eradicate Poverty. Rosenfeld, A., Dorman, M., Schwartz, J., Novack, V., Just, A.C. and Kloog, I., 2017. Estimating daily minimum, maximum, and mean near surface air temperature using hybrid satellite models across Israel. Environmental research, 159, pp.297–312.

IPCC, 2014. Summary for policymakers. Climate Change 2014: Impacts, Adaptation, and Vulnerability. Part A: Global and Sectoral Aspects. Contribution of Working Group II to the Fifth Assessment Report of the Intergovernmental Panel on Climate Change. Cambridge University Press, Cambridge, United Kingdom and New York, NY, USA,pp. 1–32.

NOAA-ESRL. 2019. https://www.co2.earth/daily-co2.

NRC (2010). Advancing the Science of Climate Changes . National Research Council. The National Academies Press, Washington, DC, USA.

USGCRP (2014). Climate Change Impacts in the United States: The Third National Climate Assessment. Melillo, Jerry M., Terese (T.C.) Richmond, and Gary W. Yohe, (Eds.) U.S. Global Change Research Program.

3

Greenhouse Gases Emissions

3.1 Chemistry of greenhouse gases (GHGs)

Greenhouse gases (GHGs) are atmospheric-gases that have capabilities to absorb and radiate-back the infrared (IR) radiation in the earth-atmosphere system (Figure 3.1). In this process the atmospheric-gases absorb and radiate-back IR radiation in lower atmosphere effectively emitting and absorbing heat energy heat energy. They maintain the heat budget of earth-atmosphere and keep our plant warm. If we typically see the chemistry of those gases, the key point is that they should have dipole moments. This is the property which makes them capable to absorb and radiate-back the range of IR radiation. In a molecule when valence-electrons are not equally shared in a bond between the two atoms, a slight positive charge is developed above one atom and a negative charge on the other atom. That charge difference of the different atoms in a

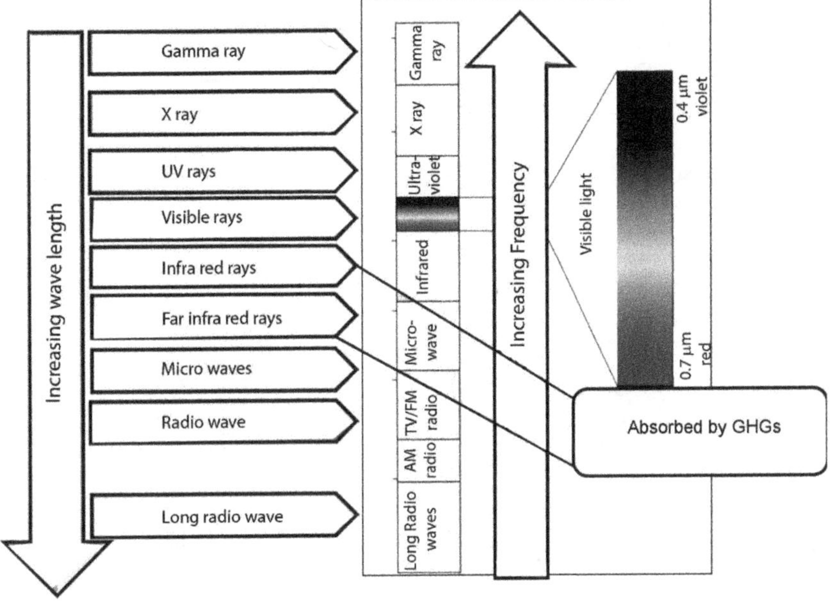

Figure 3.1 The characteristics of electromagnetic radiations showing the IR radiations absorbed by GHGs

molecule generates the dipole moment. For instance, in water vapour (H_2O-vapour), an unequal sharing of electrons and or separation of charge causes dipole moment. In H_2O, the centre has a positive charge at the hydrogen atoms and negative charges are located on the oxygen atom. The magnitude of dipole-charges, vibration and stretching of O-H bond generates certain frequency of the oscillating dipole moment that actually regulates the absorption capacity of IR radiation by H_2O vapour.

The water vapour (H_2O-vapour), carbon dioxide (CO_2), methane (CH_4), nitrous oxide (N_2O) and chlorofluro carbon (CFC) are the major GHGs in atmosphere having dipole moments. All the above mentioned gases have clear cut dipole moments which could be seen from their structural configuration except CO_2 (Figure 3.2). The carbon dioxide has symmetrical chemical bonding. So, obviously it does not have dipole moment as such. However, due to asymmetric stretching and vibration of its bonds an instantaneous dipole moment is generated that make it capable for absorbing and radiate-back the IR radiation. So, the CO_2 becomes a GHG. The other common gases in atmosphere such as, nitrogen (N=N) and oxygen (O=O), don't have dipole moments, so they are not GHGs (Figure 3.3).

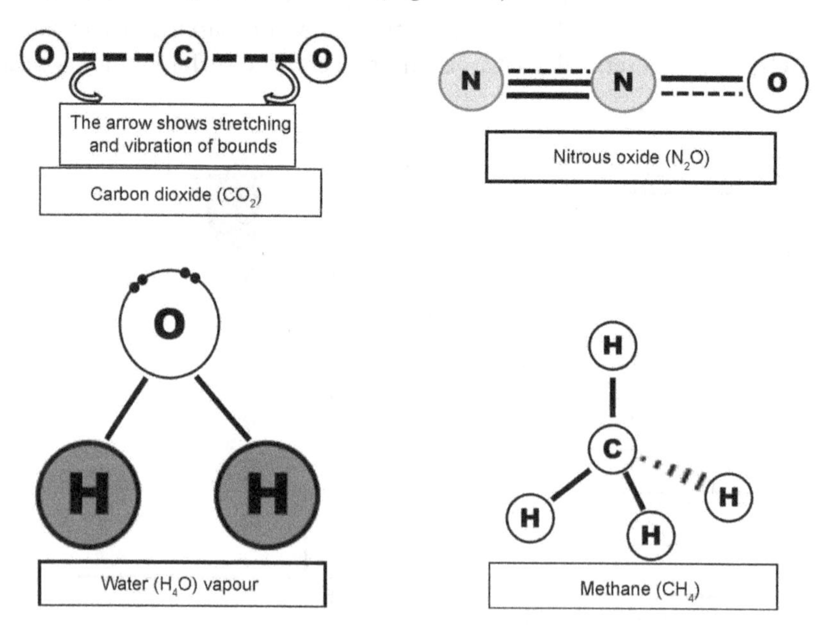

Figure 3.2 Major GHGs with the molecular structure indicating dipole moments

Hydrogen gas Nitrogen gas

Figure 3.3 Linear molecular structures without dipole moments of non-GHGs in atmosphere

3.2 Radiative forcing and energy balance

Radiative forcing (RF) refers to the net energy balance change in the earth-atmosphere system due to the imposed perturbation. The RF is expressed as Wm^{-2} (Watts per square meter) averaged over a specific period of time. The specific period of time indicates the time scale when the imposed change has been taken place on energy imbalance. The estimated RF actually provides a simple quantitative basis for comparing the potential climate-response to global mean temperature changes. It is widely used by researchers, environmentalists and even policy makers. Generally, RF is presented between two particular time scales, like, pre-industrial to recent-time. However, often different definitions of RF is used by various scientific communities. The instantaneous RF is defined as the instantaneous change in net radiative-flux including long-and short wave radiations due to imposed changes. This RF is generally expressed in terms of flux changes at troposphere.

However, the anthropogenic activities increase the concentration of the well-mixed GHGs that have substantially 'enhanced greenhouse effect'. This increases the radiative forcing causing rapid global warning. But, aerosols sometimes offset the enhancement of RF and triggers the uncertainty associated with the climate change trend. The IPCC assessments used the concept of RF long back in order to evaluate and compare the different mechanisms affecting the earth-energy balance and that causes climate change. Previously, (before AR5 of IPCC) the fixed RF of specific GHG was used in troposphere. However, presently (in AR5, IPCC, 2014) effective RF (ERF) is found more relevant due to the gradual increase of well-mixed GHGs in atmosphere. The ERF estimation actually takes into consideration all the physical variables to respond to perturbations except for those related to the ocean and sea-ice. The values of ERF and RF are found significantly different for anthropogenic aerosols because of their impact on snow cover and clouds. The RF and ERF are estimated over the Industrial Era from 1750 to 1993, 1998, 2005, 2011 (Table 3.1) and differed over time scale. The total anthropogenic ERF over the industrial-era was found 2.3 Wm^{-2}. Whatever be the positive and negative values estimated over the years, it is certain that the total anthropogenic ERF in the recent time is positive as compared to industrial-era (1750). And it has been increasing rapidly since 1970 over the prior decades. The enhancement primarily occurred due to increases in the GHGs concentration in atmosphere and the carbon dioxide's (CO_2) RF nearly increased by 10%. The emissions of CO_2 have contributed highest to anthropogenic RF in each decade since the 1960s.

Another important aspect in RF is contribution of ozone and aerosol. The tropospheric ozone that is not emitted directly to the atmosphere but is formed by photochemical reactions, largely responsible to anthropogenic emissions of CH_4, NOx, CO and non-methane VOC, while stratospheric ozone RF results primarily from ozone depletion by halocarbons (Figure 3.5). There is robust evidence that tropospheric ozone also has a detrimental impact on vegetation-physiology, and therefore on its CO_2 uptake. The magnitude of the aerosol forcing was reduced relative to AR4 (Table 3.1). Despite the large uncertainty, the aerosols have been found to offset a substantial portion of global mean forcing caused by well-mixed GHGs. Land use change also modify the net RF, particularly through the change in hydrologic cycle.

Natural forcing through solar irradiance also changed slightly since 1978 to 2011. The recent solar cycle showed a slight/minimum change and it was lower than the prior two. This has caused a very small negative RF of –0.04 Wm^{-2} between the years 1986-2008. The overall change of total solar irradiation from 1750 to 2011 was 0.05 Wm^{-2}. It is well understood that the RF of volcanic -aerosols is highest for a short period (~2 years) following volcanic eruptions.

Emission metrics such as Global Warming Potential (GWP) and Global Temperature Change Potential (GTP) could be used to quantify the relative and absolute contributions of different drivers to climate change. Those could be used for specific substances and also for the emissions from sectors, regions or countries. The metric used by policy makers is the GWP, which specifically integrates the RF of a substance over a chosen time horizon, relative to that of CO_2. On the other hand the GTP is expressed as the ratio of change in global mean surface temperature at a specific point in time from the substance of interest relative to that from CO_2. The values of RFs are significantly dependent on metric type (GWP as well as GTP) and time horizon. We have to choose the time horizon and metric type based on our objective and applications.

The RFs may be positive or negative. A negative RF cools the earth-atmosphere whereas, a positive RF warms it. The positive RFs of CO_2, CH_4, N_2O, water vapour, cholofloro carbon etc. warms our planet. The black carbon-aerosols also have a significant positive RF. Anthropogenic activities have increased the RFs of atmospheric GHGs by 27.5% (since 1990 to 2009; NOAA, 2019). The fossil fuel burning and energy sector is primarily responsible (80%) for that enhancement, and CO_2 is the major gas. The negative RFs are recorded in case of aerosols, or small particulate matters. One of the examples of that are the sulfate-aerosols that directly reflect sunlight back into space and have a negative forcing (Figure 3.5).

Table 3.1 Radiative forcing of GHGs as proposed by IPCC at different time scale.

S. No	**Radiative forcing of GHGs and different components of atmosphere**	**Global mean radiative forcing (Wm^{-2})**				**Effective radiative forcing (ERF) (Wm^{-2})**
		SAR (1750-1993)	TAR (1750-1940)	AR4 (1750-2005)	AR5 (1750-2010)	
1.	Well mixed GHGs (CO_2, CH_4, N_2O and halo carbons)	2.45	2.43	2.63	2.83	2.83
2.	Troposphere O_3	+0.40	+0.35	+0.35	+0.40	NE
3.	Aerosol-radiation interaction	NE	NE	-0.50	-0.35	-0.45
4.	Aerosol-cloud interaction	0 to1.5	0 to 2.0	-0.7	NE	-0.45
5.	Surface albedo (land use)	NE	-0.20	-0.20	-0.15	NE
6.	Black carbon and aerosol on snow and ice	NE	NE	+0.10	+0.04	NE
7.	Total anthropogenic	NE	NE	1.6	NE	2.3
8.	Solar radiation	+0.30	+0.30	+0.12	+0.05	NE

[SAR: Second Assessment Report; TAR: Third Assessment Report; AR4: IPCC Fourth Assessment Report; AR5: IPCC Fifth Assessment Report; NE: Not Estimated]
(***Source:*** WGLAR5-Chapter 8, IPCC, 2014: page 697)

Greenhouse Gases and Greenhouse Effects

The agriculturally important GHGs are carbon dioxide (CO_2), methane (CH_4) and nitrous oxide (N_2O). Their RFs are different and GWPs vary with life time and concentration in atmosphere. The comparative radiate forcing, properties of CH_4, CO_2 and N_2O and their global warming potentials are presented in Table 3.2.

There are three primary sources of GHGs in earth-atmosphere, viz., solar-sources, volcanic- sources and biospheric/ industrial- sources. Solar sources include the production of GHGs through solar proton events, changes in solar constant, and ozone content. The volcanic origin could produce GHGs after chemical conversion. Activities in biosphere, energy sector and industries are responsible for anthropogenic emission of GHGs. The industrial sectors are primarily responsible for the emission of sulphur oxides (SOx), sulphur hexafluoride, hydro-fluorocarbons, per-fluorocarbons, and the GHGs with higher GWPs. The ozone depletion is often experienced by the substances like, fluorinated- hydrocarbons (i.e., CFCs, HCFCs). Although industrial-GHGs are emitted in less quantity but as they have very high RF and GWP, should be tackled seriously in the face of climate change consequences.

The term of global warming potential (GWP) of the greenhouse gases is the metric that is very often used to quantify the effective RFs of GHGs. The GWP of greenhouse gases (Table 3.2) depends on the radiative forcing of GHGs. Actually, the radiative forcing of GHGs primarily regulate the net energy balance of earth-atmospheric system.

Table 3.2 Global warming potential and radiative properties of agriculturally important greenhouse gases in the atmosphere

GHGs	Pre-1750 concentration (ppm)	Current tropospheric concentration (ppm)	Annual increase (%)	Atmosphere life time (years)	GWP (100 years time horizons)	Increased radiative forcing (W m^{-2})
CO_2	280.0	412.0	0.5	Variable	1	1.46
CH_4	0.715	1.82	0.8	12	25	0.50
N_2O	0.270	0.325	1.0	114	298	0.15

(***Source***: IPCC, 2014; NOAA-ESRL. 2019)

Greenhouse Effects

The literal meaning of the process "greenhouse effect" is warming a house by covering it with glass which allows the light energy (mostly sunlight) to enter into it. At the same time, the glass cover primarily restricts the airflow within house and thereby makes it warm. But the 'greenhouse effect" in earth-atmosphere is different. In earth-atmosphere, the GHGs absorb and radiate back the long wave radiations (specifically IR radiation) and keep our planet warm. We must remember the surface temperature would be around -17°C or 256°K without this 'greenhouse effect'. So, obviously there would be no life-existence in this planet. So, lives sustain in our planet due to sustained 'greenhouse effect' and the recent average global surface temperature is around 14.85°C (IPCC, 2014). But we should also remember the average surface temperature in last century has been increased in the tune of 0.85°C, primarily due to "enhanced greenhouse effect" caused by anthropogenic (human) activities. The "enhanced greenhouse effect" is a matter of concern matter of concern which is not desirable for sustaining our ecosystems. The 'enhanced greenhouse effect' causes global warming. The global warming by virtue of increased temperature and associated disturbance of meteorological variable affects the agriculture and other sectors of our society both in positive and negative ways.

Global warming

The global surfaces are getting warmer at a rate higher than it is expected for a specified time scale is termed as 'global warming'. Most of the present day climate scientists think, the global warming is mainly caused by 'enhanced greenhouse effect' (Figure 3.4). The 'enhanced greenhouse effect' on earth-atmosphere is caused by increased GHGs emissions due to human activities (fossil fuel burning, deforestation, vehicular pollution, refrigeration, faulty agricultural practices, energy sector emission, etc.). Environmentalists strongly believe that the present day global warming is human-induced as climate models not satisfactorily explain the observed trends of global warming with the data-inventory of solar irradiation only. More so, the warming is only noted at lower atmosphere. The GWPs of GHGs are quantified with reference to CO_2 equivalent (keeping GWP of CO_2 '1' at a particular time scale).

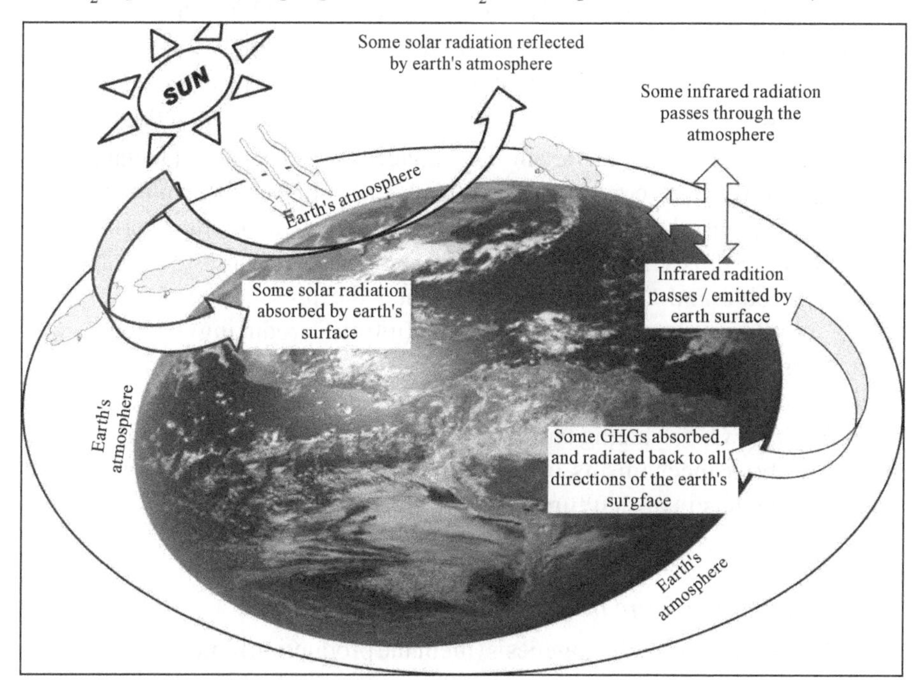

Figure 3.4 Schematic representations of greenhouse effect, enhanced greenhouse effect and global warming

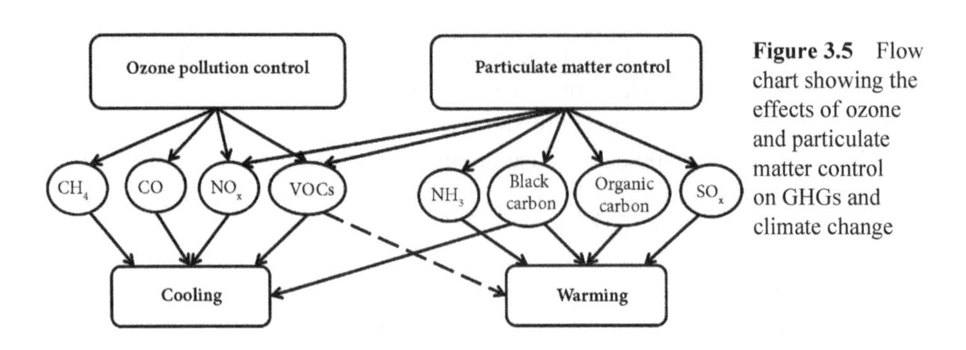

Figure 3.5 Flow chart showing the effects of ozone and particulate matter control on GHGs and climate change

3.3 Mechanism of production, transformation and emission of GHGs in agriculture

The major GHGs related to agriculture are CH_4, N_2O and CO_2. However, recent school of thought believes that the biogenic sources of CO_2 should not be considered as agricultural contribution to GHGs emission to atmosphere. But still it is a debatable issue. So, in this chapter we would discuss the mechanisms of production of all the three gases in soil, their transformation and emission from soil to atmosphere either through soil or through plants. Methane is generally produced in soils or produced through enteric fermentation in rumen of animals. The mechanisms of CH_4, N_2O and CO_2 production in soils primarily depends on three drivers. The drivers are (i) availability of quality substrate, (ii) predominant soil-conditions and (iii) presence of responsible microorganisms required for transformation of substrates to specific GHG.

The availability of sufficient carbonaceous substrates (acetate/CO_2), continuous anaerobic condition ($\leq$ 200 mV) and presence of methanogens (methane producing bacteria) are the three pre-requisite for methane production in soil. In case of N_2O, availability of nitrogenous substances (NH_4^+, NO_3^{-1}), cycle of aerobic and anaerobic condition and the presence of nitrosomonas (nitrite producing bacteria), nitrobacters (nitrate producing bacteria), and denitrifiers (nitrous oxide producing bacteria) are the drivers in soils. The CO_2 production in soil also requires carbon substrates, aerobic soil condition and heterotrophs (CO_2 producing bacteria). In the coming sections, we will discuss in details about mechanisms of production, transformation and emission of these three GHGs in agriculture.

3.3.1 Methane

Mechanism of methane formation and transformation

The balance between methanogenesis (methane production) and methanotrophy (methane consumption/oxidation) in soil actually regulate the net emission of methane in agricultural systems. The source and sink behaviors of the system depends on the relative magnitudes of those two processes. The acetate, carbon dioxide, formate, alcohols I, II and methylated compounds are the substrates used in methanogenesis in soil by methanogens in anaerobic condition. The continuous anaerobic condition with a redox potential of about around -200mV is favourable for CH_4 production. In soils' macro-pores or micro-pores this redox potential should be reached for methane production. Soil texture, structure and organic matter presence regulate this condition (redox potential around -200mV) which generally attained at 8 to 10 days of continuous submergence in fields. However, time of submergence for reaching that particular redox-condition of soil may also vary with soil chemical-microbial constitutes and site specific microclimate of soil-plant system. The reduction sequence (Table 3.3) of major elements present in soil, methane production pathways and its transformation (Figure 3.6) need to be understood to know the mechanism of CH_4 emissions in agriculture.

Table 3.3 Reduction sequence in anaerobic soil

Element (Oxidized forms)	Process	Elements (Reduced forms)	Redox potential (Eh)
Oxygen (O_2)	Reduction →	Water (H_2O)	
Nitrate (NO^{-3})	Reduction →	N_2O	> 200 mV
Manganese (Manganic form (Mn^{+4}))	Reduction →	Manganese (Manganus form (Mn^{+2}))	0 to 200 mV
Iron (Ferric form (Fe^{+3}))	Reduction →	Iron (Ferrous form (Fe^{+2}))	0 to -150 mV
(Sulphate (SO_4^{-2}))	Reduction →	Sulphide (S^{-2})	< -150 mV
(Carbon dioxide (CO_2))	Reduction →	Methane (CH_4)	< -200 mV
Hydrogen (Cationic form (H^+))	Reduction →	Hydrogen (Molecular form (H_2))	< -200 mV

Methane production in soil is a microbial mediated bio-chemical process. In microbiological terms, the process of methane production is known as methanogenesis. In this process carbonaceous compounds, preferably acetate and or CO_2 is converted to methane. Around 70-78% of the methanogenic population utilizes H_2 along with CO_2 but responsible only for 25–30% of the methane production in soil. Specifically, in lowland rice, soil methanogens mostly used acetate substrate for methane production following the acetoclastic pathways (Bhattacharyya et al., 2016). Among the methanogens, the population of Methanosarcina and Methanosaeta are relatively lesser in wetland-soil, contribute two-thirds of CH_4 production in these systems. Apart from this, root exudation (both quality and quantity), soil aggregation, water management, fertilizer schedule, and cultural practices are the other important factors which affect the methane production in rice field. The presence of elements like nitrate, iron, manganese, sulphate are also playing important role on the extent of methane production in anaerobic soil systems.

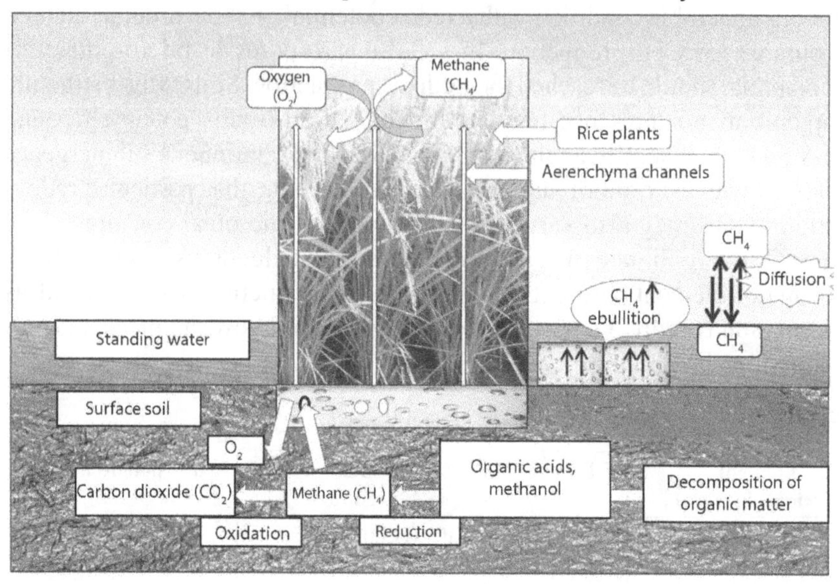

Figure 3.6 Schematic representation of mechanism of methane production and transformation in agriculture

Table 3.4 The prevalent chemical reactions in methanogenesis with equations, substrate use, predominant methanogens (genus and species level) in wetland

S. No.	Substrate used	Equations	Methanogenic groups responsible	Example of Organisms
1.	Carbon dioxide and hydrogen H_2+CO_2	$4H_2 + CO_2 \rightarrow CH_4 + 2H_2O$	Hydrogenotrophs	• *Methanobacterium bryantii* • *Methanobacterium formicum* • *Methanobacterium thermoautotrophicum* • *Methanobrevibacter ruminantium* • *Methanobrevibacter smithii*
2.	Acetate CH_3COO^-	$CH_3COO^- + H^+ \rightarrow CH_4 + CO_2$	Acetotrophs	• *Methanosarcina acetivorans* • *Methanosarcina barkeri*

S. No.	Substrate used	Equations	Methanogenic groups responsible	Example of Organisms
3.	Formate	$HCOO^- + 4H^+ \rightarrow CH_4 + 3CO_2 + 4H_2O$	Formatotrophs	• *Methanococcus aeolicus* • *Methanococcus maripaludis* • *Methanococcus vannielii* • *Methanococcus voltae*
4.	Methylated compounds		Methylotrophs	• *Methanomicrobium mobile*
5.	Alcohols I, II		Alcoholotrophs	• *Methanoculleus bourgensis* • *Methanogenium olentangyi* • *Methanogenium bourgense)* • *Methanoculleus marisnigri*

Important chemical conversions in soil during methanogenesis

Soils regulate around 60% in the atmospheric methane budget. The wetlands and anthropogenic wetlands (rice-paddy soil) are the major sources of CH_4 production in earth-atmosphere. These ecologies provide continuous anoxic conditions (limited oxygen diffusion from atmosphere to soil). On an average, about $2/3^{rd}$ of CH_4 is produced in soil through reduction of acetate and about $1/3^{rd}$ by the reduction of CO_2 and H_2 or formate (Ferry, 1992). In marine ecosystem, small quantity of methane is produced by the oxidative-reductive-dismutase pathways where, methanol or methylamines are mostly used by methanogens. Methane is also generated from dimethyl-sulfide by taking cyclic alcohols as electron donors. The prevalent chemical reactions of methanogenesis are presented in Table 3.4.

Overall, the mineralization of organic carbon in anaerobic condition i.e fermenting is done by methanogenic bacteria that produce CH_4 and CO_2. Specifically, the conversion of acetate to methane is generally carried out by *Methanothrix* and *Methanosarcina,* whereas, the CO_2 reduction is primarily done by *Methanobacterium.* In the bio-chemical front, the hydrolysis of biological polymers into monomers governed by hydrolytic microflora which are usually obligate (strictly) anaerobes. However, monomeric compounds and intermediary compounds (volatile fatty acids, organic acids, alcohols) formed during acidogenesis by fermentative microflora which are either facultatively or strictly anaerobic.

Methanotrophy

The methanotrophy (the oxidation of methane to CO_2) is generally discussed separately but it is the integral part of methanogenesis. In methanotrophy, conversion of methyl group to CO_2 occurs and O_2 acts as electron acceptor. So, in simple words the bacteria which utilize CH_4 are referred to as methanotrophs. The majority of these bacteria use molecular oxygen as the terminal electron acceptor and therefore grouped as obligate aerobes. But few are also reported to habitat in SO_4^{-2} reducing anaerobic environment and oxidise CH_4.

The monooxygenase enzyme is responsible for the first step of CH_4 oxidation. It needs molecular O_2. In between, the methanol is produced which is further oxidized to CO_2 via formaldehyde and formate. The catalytic nature of methane monooxygenases is the unique feature for oxidation of CH_4 to methanol by methanotrophs. However, methane monooxygenases present in aerobic methanotrophic bacteria have lack of substrate specificity that results in several intermediate compounds including xenobiotics.

Interestingly, the CH_4 is also consumed in other anaerobic environments, including anoxic marine water, sediments of soda lakes and freshwater sediments. In vertical profiles of marine sediments, methane oxidation and sulphate reduction occur simultaneously. The sulphate dependent methane oxidation has also been accepted to occur in nature. Generally, the anaerobic CH_4 oxidation is determined by tracer technique by measuring the conversion of $^{14}CH_4$ to $^{14}CO_2$.

In this connection, it is need to be known that in rice soil, 58-80% of locally produced CH_4 is oxidized there itself. And more so, the quantum of CH_4 oxidation is usually higher in rice-planted soil than fallow. The oxidation rates of methane is also varying with rice growing stages; *viz.*, about 37 and 55% of CH_4 in rice-soil is oxidized at tillering and panicle initiation stages, respectively. In harvesting and ripening stage, the oxidation rates of methane in rice soil are negligible. In nut shell, the methane oxidation rates in rice soil primarily depends on the oxidation power of roots that estimated through oxidation of a-naphthylamine which decreases with the age of the crop.

Drivers of methane production and transformation

There are five major drivers of methane production and transformation in agricultural soils. Those are water regime, soil amendments, root exudation, soil labile carbon pools and active methanogenic population.

Saturated water regime in soil creates anaerobiosis and that favours the CH_4 production. Flooding triggers CH_4 production and has strong influence on its transmission as well as emission rates in rice-paddy. The alternate wetting and drying moisture regime reduces methane production. The soil amendments like, *dhaincha* (*Sesbania;* green manuuring crop), azolla and compost increased the CH_4 production, whereas, phosphogypsum, basic slag were reported to reduce the production of methane in soil. The *Sesbania*, azolla and compost (organic matter amendments) enhances the readily mineralizable soil organic carbons which are the food source of methanogens (methane production in flooded soil). On other hand, sulphur and silicon present in phosphogypsum and basic slag, respectively compete with carbonaceous substance present in soil for electron transport and thereby reduces the CH_4 production. The third important driver is root exudation. The easily decomposable organic substrate secreted through root exudates favour CH_4 production in soil. These easily decomposable organic substances secreted through root exudation are good food/energy sources for CH_4 producing organisms (methanogens). However, root exudation varies with genotypic and phenotypic variations of crops. Even within the same crop the quantity and quality of exudation are varied with stages of crop growth. The exudation rates are usually lower at seedling stage and increase up to flowering in rice then again decrease at maturity. And interestingly, similar trends of methane emission were noticed in rice with their growth stages. The large variation in the quantity (exudation rate) and quality (composition) of root exudates in different cultivars of rice (Oryza *sativa)* were reported by Bhattacharyya et al., (2013). The rice roots exudates higher amount of malic acids followed by tartaric, succinic, citric and lactic acids (Aulakh et al., 2001). Methane production and emission are more closely related to the release pattern of root exudates-C than its individual components. The proportion of exudates-C converted to CH_4 ranged between 61 and 83% (Figure 3.7) (Aulakh et al., 2001)

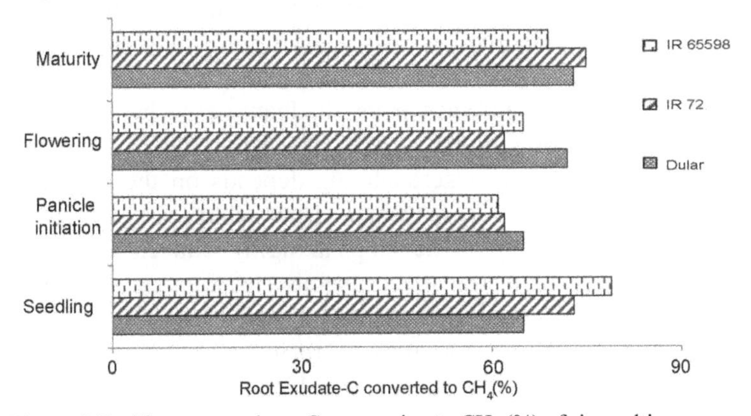

Figure 3.7 The root exudates-C conversion to CH_4 (%) of rice cultivars (IR65598, IR72 and Dular).

The last but not least, two drivers of methane production in soils are soil labile carbon pools and methanogens activities. In simple words, the higher labile

C-pools in soil stimulate methane production by supplying food and energy to methanogens.

Mechanism of CH_4 emission from soil

There are three main processes by which methane is emitted from soil to atmosphere. Those are diffusion, ebullition and through plant-conduit system. The diffusion of methane takes place due to concentration gradients of methane between soil-air and atmospheric-air. If the concentration of methane is more in soil-air than in atmosphere-air, then methane is emitted from soil to atmosphere and vice-versa. In the ebullition process, methane escapes from soil though water-bubble primarily from wet land and rice-paddy soil. And the third process, i.e through plant-conduit system, methane is escaped around 85-90% from rice fields. Therefore, in rice-paddy system, the plants serve as the major conduit for the transfer of CH_4 from soil to atmosphere. However, the CH_4 transport capacity differs greatly among rice cultivars.

The mechanisms of CH_4 transport/emission through rice-aerenchyma tissues

Methane is not produced in rice plant. Methane is produced in soil. Rice plant only provides passage to escape it from soil to atmosphere. So, rice should not be held guilty for CH_4 emission. And more so, the conduit (capillary-tube-spaces) that is provided by rice palnt for methane emission is not made for that. The main function of aerenchyma formation in rice (hydrophilic plants), is to deliver oxygen to its roots. But unfortunately, many gases including methane are transferred through them into the opposite direction. Actually, aerenchyma in rice is made up of lacunae with varying sizes (like small, medium and large). The density and size of aerenchyma-lacunae significantly control the CH_4 transport capacity of rice. Further, the aerenchyma formation in rice is a varietal character. For instance, in cultivar-Varshadhan; aerenchyma spaces are well developed whereas, in Kalinga 1; it is poorly formed (Figure 3.8). The development and orientation of aerenchyma depends on the stages of crop growth and redox-condition of rhizosphere (Bhattacharyya et al., 2019). Generally, 'wide-oriented aerenchyma' develops in highly reduced conditions in rhizosphere. In rice, the methane transport capacity was found maximum at panicle initiation stage followed by maximum tillering, active tillering, maturity and seedling stage. Low methane transport at maturity stage due to collapse of 'wide-aerenchyma lacunae' and subsequent blockage of aerenchyma made capillary channels. In nut shell, the process of CH_4 transport/ emission through rice plant involves four definite steps *viz.*, (i) diffusion of CH_4 into the root from soil, (ii) conversion of CH_4 at root cortex, (iii) diffusion and upward movement of CH_4 through cortex and aerenchyma-capillary and (iv) emission to the atmosphere through micro-pores of leaf sheath. The entry of methane to root cortex is usually taken place through the cracks in the junction point of the main root and or through root hairs. However, root can also absorb the gaseous from of methane. And finally, methane is released to the atmosphere primarily through micro-pores in the leaf sheaths in the lower leaves but not through stomata. Majority of the CH_4 released is chanelled through the culm. Around 50-60% of the CH_4 is released through leaf blades before and during shoot elongation, and only a little fraction through leaves. Methane could also be emitted through panicles during submergence when the vegetative parts are drowned below the water level.

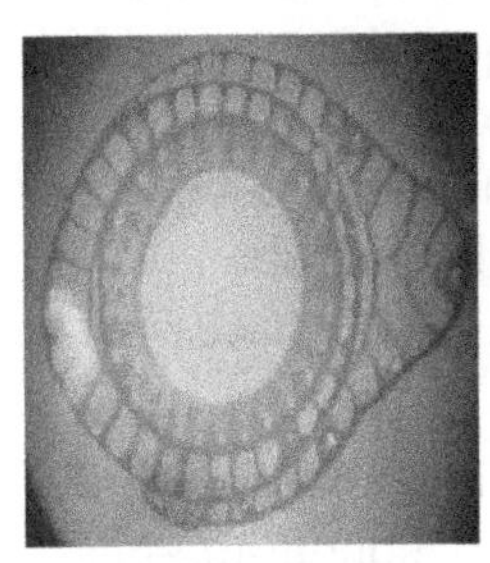

Kalinga 1

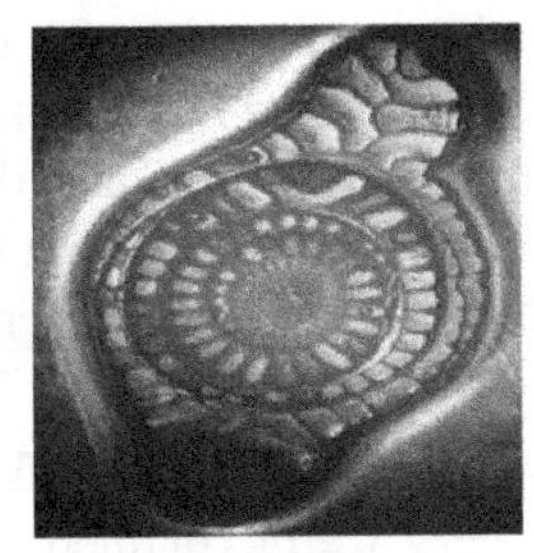

Varshadhan

Figure 3.8 Aerenchyma orientation of rice cultivars Kalinga 1 and Varshadhan.

Here, one point can be noted that the wide variations among rice cultivars with regard to CH_4 flux, opens up a great opportunity to breed low CH_4 emitting rice varieties. However, in rice, not only capillary conduit formed by aerenchyma tissue regulates the methane emission but also the oxidative capacity of root plays an important role. Actually, rice influences the oxidation of CH_4 by two ways. First process is the diffusion of atmospheric O_2 into the rhizosphere through aerenchyma, and the second one by enzymatic oxidation at root surface. The oxidative capacity of rice roots is generally measured by N-flush inhibition technique and or by a-naphthylamine oxidation method. So, the rice cultivars having higher root-oxidative capacity should be considered for mitigating CH_4 emission.

3.3.2 Nitrous oxide

Mechanisms of N_2O production and transformation in soil

Nitrous oxide is produced in soil through an oxidation-reduction cyclic process. At first nitrate is produced from ammonia/ammonium by oxidation. This oxidation process is also called nitrification, where at first step nitrite is formed by nitrosomonas bacteria and then nitrate is generated from nitrites by nitrobacter. The reduction process is referred as denitrification where in anaerobic condition nitrous oxide is produced from nitrates by denitrifiers.

So, nitrification occurs in aerobic conditions, whereas, denitrification prevails in anaerobic conditions. During denitrification along with N_2O the dinitrogen (N_2) or nitrogen oxides (NO) also could be produced. In the first step of denitrification NO is formed. And in continuous anaerobic condition, considerable amount NO is further reduced to N_2O or N_2 by denitrifiers.

Now, we will discuss in details the nitrous oxide production and transformation in soil system (Figure 3.9). The nitrification is the oxidation process in which ammonium is transformed to nitrite and subsequently to nitrate. Nitrosomonas and Nitrobacters are the two bacterial groups responsible for nitrification. However, the first step of nitrification is very fast and intermediate product that is nitrite is very unstable. So, quantification of first step of nitrification is not easy. The product of nitrification is the nitrogenous-substrate required for the denitrification process and thereby production of N_2O. In denitrification process, a series of pathways for the formation of N_2O via the intermediate compounds NH_2OH or NO has been taking place (Naqvi and Noronha, 1991) (Figure 3.10).

The regulative mechanism of nitrification and denitrification and subsequent nitrous oxide formation could be very well explained by "hole-in-the-pipe," theory (Firestone and Davidson, 1989).

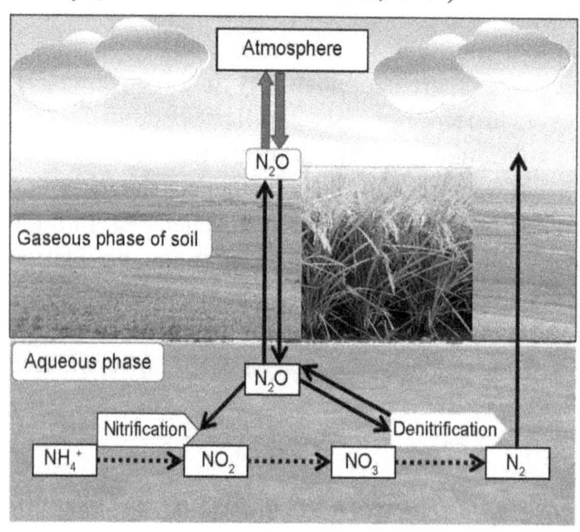

Figure 3.9 Schematic representation of mechanism of nitrogen oxide production and transformation in agriculture.

Mechanism of (N_2O) production in soil system

The "hole-in-the-pipe" theory suggests two levels of control that regulate emissions of N_2O and NO from soil. The nitrogen flow depends on the ratio of N_2O: NO. The factors that regulate that ratio are called the 'leakage' of the pipe. And the "hole-size" (that caused leakage) of the pipe depends on the relative influence of the factors. The bigger "hole-size' refers to the greater influence of that particular factor on regulation the N_2O: NO ratio in the system (here described as pipe). As for the example, in well-fertile soils, flow through the pipe is more, and also the "leaks." The opposite is true for infertile soils. The nitrification "leak" is higher in dry soils, as oxygen is present. So, in dry soil, the production of NO would be relatively more than N_2O and N_2. In wetter soils (anaerobic condition), denitrification is dominant and relatively more N_2O is produced. Further, in continuous wet soils, denitrification process results higher reduced form of nitrogen (i.e di-nitrogen; N_2) as the end product (Davidson *et al.*, 2000).

Drivers of N_2O production and transformation in soil

There are five major drivers of N_2O production and transformation in soil. Those are type and dose of N-fertilizer, soil moisture regimes, active nitrifier and denitrifier population, soil organic matter contents and soil-physicochemical properties.

Type and dose of nitrogenous fertilizer regulates the N_2O production. Higher N dose in general, provides more N-substrates and causes more N_2O production. The coated urea and urea super granules/ briquettes control nitrate transformation, hence reduces N_2O production and emission (Majumdar *et al.*,2000). Like methane, soil moisture regimes also influence the nitrous oxide production and emission. It regulates the transport of O_2 into soil and NO/N_2O/ N_2 out of the soil systems. The water filled pore space (WFPS) significantly controls the N_2O: NO flux ratio in soil (Davidson *et al.*, 2000). The nitrosomonas and nitrobacter population is very important for nitrification. It is well distributed in aerobic soil, however, in rice-paddy system its active population is disturbed due to continuous submergence and breaking of soil aggregates. The fourth but important driver of N_2O production and transformation in soil is, nature and amount of organic matter in soil. The relative abundance of electron donors (soil labile carbon pools) and acceptors (nitrate) affects the relative proportions of N_2, N_2O, and NO formation and emission in soil systems (Firestone and Davidson 1989). The readily metabolized organic carbon and the availability of water soluble organic matter are the major controlling factors. Not only the organic matter the physiochemical properties of soil like, pH, texture and salinity influence the nitrification and denitrification processes and subsequently production and emission of N_2O. As for example, in coarse-textured soils N_2O emission is six times higher than heavy-textured soils. Similarly, the N_2O production lasted for 3, 10 and 21 days in loamy, silty and sandy soils, respectively. Salinity inhibits both denitrification and nitrification (Inubushi *et al.*, 1999). The very important point is to be remembered here, that the nitrification is sensitive to extremes in soil pH. The pH range of 7 to 8 is suitable for nitrification (Haynes, 1986).

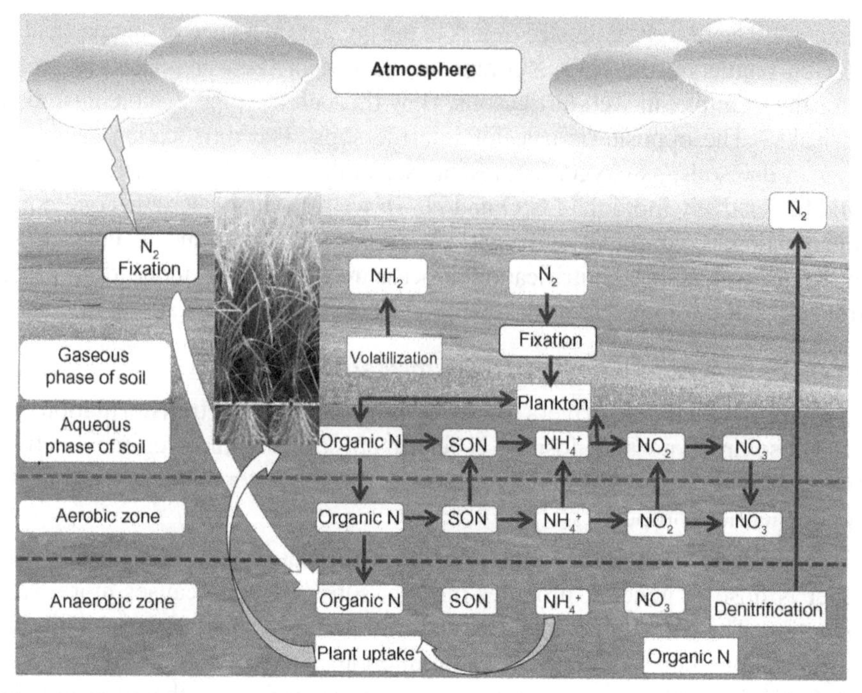

Figure 3.10 Nitrogen transformation in rice soil

Note: SON indicates soluble organic N

3.3.3 Carbon dioxide

Mechanisms of CO_2 production and transformation in soil

The net ecosystem CO_2 exchange between the soil and atmosphere is the balance between CO_2 assimilation by photosynthesis and ecosystem respiration (RE). The respiratory fluxes include respiration of autotrophs (mainly roots) and heterotrophs (microbes in soil and rhizosphere). Therefore, soil respiration includes root and microbial respiration, and basic decomposition of soil organic carbon (Hill et al., 2004). The mechanisms of respiration is well known and mostly controlled by metabolic activities of roots and microbes. Discussion of those basic physiological mechanisms is out of the scope of this book and not necessary as already known and taught at school levels. So in this section, we will discuss the drivers of CO_2 production, transformation and emission in soil-atmosphere systems and their practical consequences.

There are five major drivers of CO_2 formation in soils. Those are, soil properties (pH, soil temperature, soil aeration), moisture regimes, tillage practices (soil aggregation and compactness), soil labile C:N ratio and activities of heterotrophic population. The soil temperature and respiration have positive correlation up to 30–35°C depending on soil moisture status. Optimum pH (6-8) for heterotrphs is favorable for respiration. Soil structure/aggregation is strongly correlated with the organic carbon content and microbial populations. Soil organic carbon stabilizes the soil aggregates and at the same time aggregates protect carbon for further decomposition. Soil aggregates act as absorber for soil moisture. In general, intensive tillage and continuous cropping leads to decline in soil-aggregates and causes higher decomposition of SOC which leads to emission of CO_2 to atmosphere. Specifically, intensive tillage could increase CO_2 emission from soil by disrupting soil aggregates, incorporating plant residue and facilitating oxidation of SOC (Zhang et al., 2011). Soil macro-aggregates help in protecting labile carbon and encouraging C sequestration (Manna et al., 2005). Tillage favours higher contact between soil and C-residues and enhances soil temperature, and by the way facilitates organic matter decay and increase the soil CO_2 emission (Lal, 2004). The labile C:N ratio significantly influence CO_2 production, transformation and emission. Addition of manure and fertilizer also increases labile-C and N pools in soil and thereby (Zhang et al., 2009) enhance higher CO_2 emission to atmosphere. Even any input of N to soil could affect the C accumulation and distribution within the soil-microbes-plant systems (Coulter et al., 2009). As usual, like other GHGs emission from soil, the CO_2 emission is also directly regulated by soil moisture regimes. Alternate wetting and drying and aerobic cultivation favours carbon dioxide production and emissions from agriculture as compared to flooded rice cultivation. Last but not the least, active heterotrophic population must be maintained in soil for optimum soil respiration and consequent nutrient availability to plants.

3.4 Contribution of agriculture to GHGs emissions

The exact contribution of natural and anthropogenic activities to GHGs emission to earth-atmosphere is difficult to measure. As both are interrelated and depend on each other. However, we can give an estimation based on previous available data-trend and up-scaling through simulation modeling. Moreover, the sector wise contribution to emission is further complicated. As for example, the emission for fertilizer production should be considered under industry or agriculture is a debatable issue. Similarly, emissions of GHGs for the production of industrial raw material (like sugarcane) should be considered under agricultural head or industrial head is a question. Therefore, although we would be discussing the sector wise emission in this chapter, but proper GHGs budgeting should be done on the basis of life cycle analysis.

The quantification of agricultural contribution to gross or net GHGs emissions is difficult due to three reasons. First, separating out of the anthropogenic component of net GHGs emissions from agriculture to the atmosphere is tough. Secondly, uncertainty of available input data-inventories both state and country level and lastly, existing up-scaling models often contradicts with each other.

GHGs emission from Agriculture, Forestry and Land Use Change (AFOLU)

The IPCC reports are still considered as the most reliable and precise one in case of GHGs emission and climate change consequences. The 'agriculture, forestry and other land use' (AFOLU) was considered together in fifth assessment report of IPCC (2014) for assessing GHGs emission in this sector. Overall, this sector contributes 10-12 Gt CO_2 eq yr^{-1} of GHGs emission globally. And the major causes listed are deforestation, emission from livestock, soil and nutrient management. This sector contributes around 1/4th of total anthropogenic GHGs emission globally. The core agricultural activities contribute around 60%, where as LUC is responsible for 40% of total emission in this sector. In last four decades, forestry and land-use changes contributed around 37-48% (Figure 3.11), followed by draining of peat soils contributing a considerable portion 12-15%. Now, among the core agricultural activities, enteric fermentation of rumen contributed maximum (18-27%), followed by rice-paddy and use of synthetic fertilizer (4-6 and 3-4%, respectively) (Figure 3.12). However, one encouraging fact is that, the contribution of AFLOU sector to average GHGs emission globally decreased by 24% during the 2000-2009 over the base year 1990-2000 (IPCC, 2014). The primary reasons for that are, intervention of mitigation technologies, declining rates of agricultural area expansion, and also increase in emission in the energy sector (IPCC, 2014).

The non-CO_2 GHGs emission is considerable in agricultural sector. It was about 5.2-5.8 Gt CO_2 eq yr^{-1} (during 2010; FAOSTAT 2013) and accounted for 10-12% of global anthropogenic emissions. Additionally, the uses of farm-machineries like, tractor, power tiller, combined harvester, seed drills, irrigation pumps etc also contributed around 0.4-0.6 Gt CO_2 eq yr^{-1} (in 2010) (FAOSTAT, 2013).

The drainage of peat/ bog soil and conversion of wet lands to agriculture results in a rapid increase of GHGs emissions from soil to atmosphere. Deforestation of mangroves (a huge sink of carbon) causes a huge loss of carbon storage, biodiversity and ecological services along with emission of GHGs. Recent studies showed that in last decade the GHGs emission due to deforestation of mangrove was at the tune of 0.07 to 0.42 Gt CO_2 eq yr^{-1} (Donato et al., 2011).

So, by and large the agricultural sector collectively contributes ~23% of the GHG emissions. And the major processes/practices/systems includes decomposition of SOC (mainly emission of CO_2), rice-paddy cultivation (mainly emission of CH_4), enteric fermentation in rumens (mainly emission of CH_4), synthetic fertilizer application (mainly emission of N_2O) (IPCC, 2001, 2007, 2014; WG, 2019). The rate of GHGs emission increased dramatically during last 100 years particularly after the introduction of synthetic fertilizers and conversion of wetlands. Specifically, the annual rate of increase of GHGs from agriculture increased @ 1.6% yr^{-1} from 1961 to 2010. Overall, agriculture contributes around 0.04 Gt of CO_2, 3.3 Gt of CH_4 and 2.8 Gt of N_2O and in terms of CO_2 equivalent (IPCC, 2014) annually. But the point of concern is that around ½ of the global CH_4 and N_2O (having higher GWP) emissions has been from this sector (Figure 3.12).

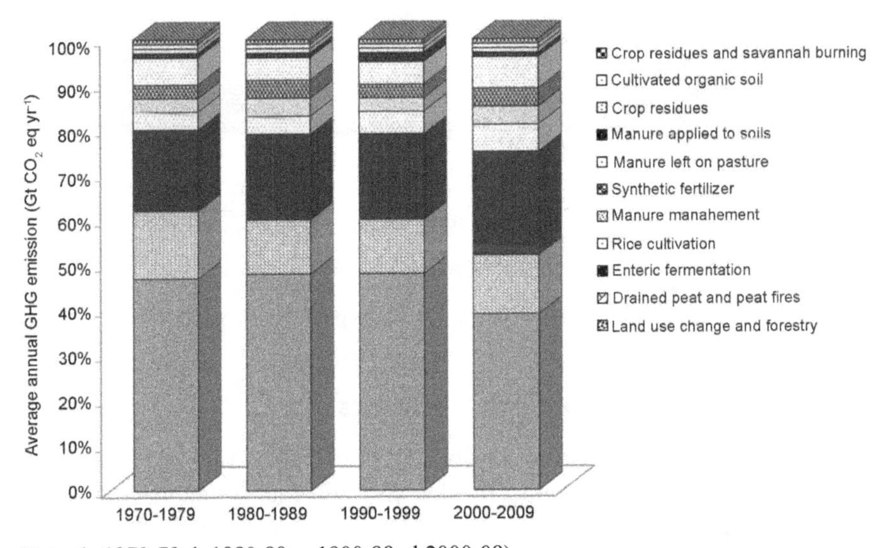

Note: (a:1970-79; b:1980-89; c:1900-99; d:2000-09)

Figure 3.11 Global trends in total GHGs emission from agriculture, forestry and land use sector over four decades

Rice-paddy cultivation is an important source of global CH_4 (11%) emissions after enteric fermentation from livestock (rumen-animals). The emissions of enteric fermentation increased from 1.4 to 2.1 Gt CO_2 eq yr^{-1} in last four decades (1961 to 2010; FAOSTAT, 2013). Losses of nutrients as well as emission of GHGs are much higher from manure deposits than manure applied to soil. However, according to recent trends, synthetic N-fertilizers are becoming a higher source of emissions than manure applied or deposited on pasture. It is now the second largest among the different components of agricultural emission categories just after enteric fermentation.

In summary, Agriculture, Forestry and Other Land Use (AFOLU) activities accounted for around 13% of CO_2, 44% of methane (CH_4), and 82% of nitrous oxide (N_2O) emissions from human activities globally during 2007-2016, representing 23% (12.0 +/- 3.0 Gt CO_2e yr^{-1}) of total net anthropogenic emissions of GHGs. However, the human-induced environmental changes caused a net sink of ~11.2 Gt CO_2 yr^{-1} during 2007-2016 (equivalent to 29% of total CO_2 emissions). Now, if emissions associated with pre- and post-production activities in the global food systems are included, then the emissions would be approximately 21-37% of total net anthropogenic GHG emissions.

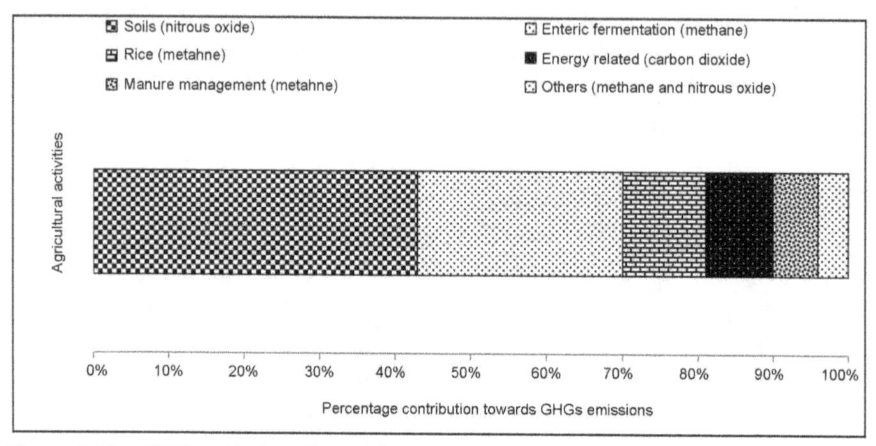

Figure 3.12 GHGs emissions from different components/practices in Agriculture

3.5 GHGs emission from other sectors

The 85–90% of anthropogenic CO_2 emissions in earth-atmosphere is from fossil-fuel burning. This is primarily due to high energy-intensive activity throughout the world including power generation, materials-industry production and transport services. As for example, the CO_2 emission enhanced by around 2.3-5% in last couple of years in the developing countries like India, Brazil and Indonesia. However, the point to be remembered that the developed countries (e.g., (including China) United States and European Union) contributing around 50-60% of total global CO_2 emission (~19-20 billion ton CO_2 equivalent). The IPCC (2014) listed the six highest CO_2 emitting countries/regions in world as United States, European Union, Russian Federation, China, Japan and India. The oil products consumption in transport sector, refineries, energy consumption in manufacturing and building industries have been showing an increasing trend over the last decades with an overall increase in the range of 0.5-2%. However, biofuel utilization in transport-sector has also been enhanced by around 7-8% during that time (IPCC, 2014).

India's CO_2 emissions although increasing over last decades but per capita emissions were much lower than those of developed countries and China.

For instance, the increase in coal consumption in India was around 59% of total fossil fuel consumption during 2013. In steel industry the CO_2 is emitted primarily during iron and steel manufacturing processes when coke ovens, blast furnaces and basic oxygen-steel furnaces are used. In cement industries, carbonate oxidation is the largest of non-combustion sources of CO_2 from industrial sector. It contributes around 5-9% of CO_2 emission to atmosphere globally, varies with regions and countries.

We are fortunate and thanks to contributions of scientist and technocrats that in recent years renewable energy sources meet almost 1/5th of global energy consumption (REN21, 2014; IPCC 2014). Recent data reveals that the renewable energy sources contribute around 56-60% of the electricity production globally. Those include hydropower (16.4%), wind power (2.9%) and biomass-power (1.8%). Further, presently, at least 144 countries, 2/3rd of which are developing countries, have their own renewable energy targets. Moreover, the total global solar power capacity has also been increased by 39% to about 138.9 GW (during 2013), and expected to produce ~ 160 terawatt hours (TWh) of electricity globally every year (IPCC, 2014).

3.6 Key messages

1. The GHGs are those gases in atmosphere which have dipole moment and could absorb and radiate-back the long wave radiations (IR).
2. Greenhouse effect makes our planet warm but 'enhanced greenhouse effect' caused by increased GHGs concentration in atmosphere causes global warming.
3. Global warming potential depends on radiative forcing of individual GHG.
4. The balance between methanogenesis and methanotrophy determines the net methane emission from agriculture.
5. Methane is not produced in rice. It is produced in soil. Rice only provides conduit to escape it from soil to atmosphere.
6. Three basic things required for methane production in soil, namely, labile carbonaceous substrates, strictly anaerobic condition (-200mV), and active methanogen population.
7. There is trade-off between methane and nitrous oxide production in soil-system. Methane requires strictly anaerobic condition, whereas, nitrous oxide production demands alternate aerobic and anaerobic condition.
8. Agricultural sector along with forest and land-use-change contributes around 23% of total anthropogenic GHGs emission in earth-atmosphere system.

9. Among agricultural activities/ processes the enteric fermentation in rumen contributes maximum GHGs emission.
10. Agriculture is the prime cause of anthropogenic N_2O emission from soil to atmosphere.

3.7 Probable questions

1. Explain the greenhouse gases and green house effect.
2. Write important characteristics of GHGs.
3. What is global warming and global warming potential of GHGs?
4. Write the major sources of GHGs in atmosphere.
5. Explain relative contribution of different sectors of Agriculture, Forestry and Land Use (AFOLU) to GHGs emission in earth-atmosphere system.
6. What is the contribution AFOLU to GHG emission over the last four decades?
7. Explain the mechanism of methane production and transformations in soil?
8. What are the drivers of methane production, transformation and emission from soil to atmosphere?
9. Briefly describe the mechanisms of nitrous oxide production and transportation in soil with diagram.
10. What are the drivers/ factors affecting the N_2O production and emission in agriculture?
11. Write the three basic things required for methane production in soil.
12. Differentiate methanogenesis and methanotrophy?
13. Rice is not guilty for methane emission – Justify the statement with logic.
14. Briefly describe the major microorganisms involved in methanogenesis with their substrate requirement.
15. Write the mechanism of methane transport through rice plant?
16. What are the major drivers of CO_2 emission in agriculture?

References

Aulakh. 2001. Impact of root exudates of different cultivars and plant development stages of rice (Oryza sativa L.) on methane production in a paddy soil. Plant and Soil. 230, 77–86.

Bhatia, A., Pathak, H., & Aggarwal, P.K. 2004. Inventory of methane and nitrous oxide emissions from agriculture soils of India and their global warming potential. Current Science, 87, 317–324.

Bhattacharyya, P., Nayak, A. K., Mohanty, S., Tripathi, R., Shahid, Md., Kumar, A., Raja, R., Panda, B. B., Roy, K. S., Neogi, S., Dash, P. K., Shukla, A. K., Rao, K.S. 2013a. Greenhouse gas emission in relation to labile soil C, N pools and functional microbial diversity as influenced by 39 years long-term fertilizer management in tropical rice. Soil and Tillage Research. 129, 93–105.

Bhattacharyya, P., K.S. Roy, M. Das, S. Ray, D. Balachandar, S. Karthikeyan, A. R. Nayak and T. Mohapatra. "Elucidation of rice rhizosphere meta-genome in relation to methane and nitrogen metabolism under elevated carbon dioxide and temperature using whole genome metagenomic approach." Science of the total environment 542 (2016): 886–898.

Bhattacharyya, P., Dash, P. K., Swain, C.K., Padhy, S.R., Roy, K.S., Negi, S., Berliner, J., Adak, T., Pokhare, S.S., Baig, M.J. and Mahapatra, T., 2019. Mechanism of plant mediated methane emission in tropical low land rice. Science of the Total Environment, 651, pp. 84–92.

Coulter, J.A., Nafziger, E.D., Wander, M.M., 2009. Soil organic matter response to cropping system and nitrogen fertilization. Agron. J. 101 (3), 592–599.

Davidson EA, Keller M, Erickson HE, Verchot LV, Veldkamp E. 2000. Testing a conceptual model of soil emissions of nitrous and nitric oxides. Bioscience. 50(8):667–680.

Donato D.C., J.B. Kauffman, D. Murdiyarso, S. Kurnianto, M. Stidham, and M. Kanninen. 2011.. Mangroves among the most carbon-rich forests in the tropics. Nature Geoscience. 4, 293–297. doi: 10.1038 / ngeo.1123, ISSN. 1752–0894.

FAOSTAT (2013). FAOSTAT database. Food and Agriculture Organization of the United Nations. Available at: http://faostat.fao.org/.

Ferry, J.G. 1992. Biochemistry of methanogenesis. Critical Reviews in Biochemistry and Molecular Biology. 27, 473–503.

Firestone M.K, Davidson E.A. 1989. Microbiological basis of NO and N_2O production and consumption in soil. In Exchange of Trace Gases between Terrestrial Ecosystems and the Atmosphere., Andreae M.O, Schimel D.S, (eds), John Wiley & Sons, New York, pp 7–21

Haynes R J. 1986. Nitrification. In: Haynes R J (ed.) Mineral Nitrogen in the Plant-Soil System. John Wiley & Sons,Manchester, UK. 127–165.

Hill, P.W., Marshall, C., Harmens, H., Jones, D.L., & Farrar, J. 2004. Carbon sequestration: Do N inputs and elevated atmospheric CO_2 alter soil solution chemistry and respiratory C losses? Water Air and Soil Pollution. 4, 177–186.

Inubushi, K., Barahona, M. A., & Yamakawa, K. 1999. Effects of salts and moisture content on N2O emission and nitrogen dynamics in Yellow soil and Andosol in model experiments. Biology and Fertility of Soils, 29(4), 401–407.

IPCC (2001). Summary for Policy Makers. In: Climate Change 2001: Impacts Adaptation and Vulnerability. Contribution of Working Group I to the Third Assessment Report of the Intergovernmental Panel on Climate Change

[J. J. McCarthy, O. F. Canziani, N. A. Leary, D. J. Dokken, K. S. White (eds.)]. Cambridge University Press, Cambridge, United Kingdom and New York, NY, USA. Available at: http://ipcc.ch/publications_and_data/publications_and_data_reports.shtml.

IPCC 2007. The Physical Science Basis. In: Solomon, S, Qin D, Manning M, Chen Z, Marquis M, Averyt KB, Tignor M, Miller HL (eds.) Climate Change 2007: Contribution of Working Group I to the Fourth Assessment Report of the Intergovernmental Panel on Climate Change, Cambridge University Press, Cambridge, United Kingdom and New York, NY, USA.

IPCC, 2014: Climate Change 2014: Impacts, Adaptation, and Vulnerability. Part A: Global and Sectoral Aspects. Contribution of Working Group II to the Fifth Assessment Report of the Intergovernmental Panel on Climate Change [Field, C.B., V.R. Barros, D.J. Dokken, K.J. Mach, M.D. Mastrandrea, T.E. Bilir, M. Chatterjee, K.L. Ebi, Y.O. Estrada, R.C. Genova, B. Girma, E.S. Kissel, A.N. Levy, S. MacCracken, P.R. Mastrandrea, and L.L. White (eds.)]. Cambridge University Press, Cambridge, United Kingdom and New York, NY, USA, 1132 pp.

Lal, R. 2004. Soil carbon sequestration impacts on global climate change and food security. Science. 304, 1623–1627.

Luyssaert S., E.-D. Schulze, A. Börner, A. Knohl, D. Hessenmöller, B.E. Law, P. Ciais, and J. Grace. 2008.. Old-growth forests as global carbon sinks. Nature. 455, 213 – 215. doi: 10.1038 / nature. 07276, ISSN: 0028-0836.

Majumdar, D., Kumar, S., Pathak, H., Jain, M. C., & Kumar, U. 2000. Reducing nitrous oxide emission from an irrigated rice field of North India with nitrification inhibitors. Agriculture, ecosystems & environment, 81(3), 163–169.

Manna, M. C., Swarup, A., Wanjari, R. H., Ravankar, H. N., Mishra, B., Saha, M. N. 2005. Long-term effect of fertilizer and manure application on soil organic carbon storage, soil quality and yield sustainability under sub-humid and semi-arid tropical India. Field Crop Res. 93, 264–280.

NOAA: National centers for Environmental information, climate at a Glance: Global Time Series, Published June 2014, retrieved on July 3, 2014 from http://www.ncdc.noaa.gov/cagt.

Naqvi, S.W.A., Noronha, R. J., 1991. Nitrous oxide in the Arabian Sea. Deep Sea Research. 38, 871–890.

REN21 (2013). Renewables 2013 Global Status Report. REN21 Secretariat, Paris, France, 178 pp. Available at: http: //www.ren21.Net/ren21activities/globalstatusreport. aspx.

Zhang, H., Wang, X., Feng, Z., Pang, J., Lu, F., Ouyang, Z., Zheng, H., Liu, W., Hui, D. 2011. Soil temperature and moisture sensitivities of soil CO2 efflux before and after tillage in a wheat field of Loess Plateau, China. J Environ Sci. 23, 79–86.

Zhang, W., Xu, M., Wang, B., Wang, X.J., 2009. Soil organic carbon, total nitrogen and grain yields under long-term fertilizations in the upland red soil of southern China. Nutrient Cycling in Agroecosystems. 84, 59–69.

4

Techniques of Measurement of GHGs

The measurement of greenhouse gases (GHGs) has been started during middle of the last century. Initially, CO_2 and water vapour were measured with other common meteorological variables like temperature, sunshine hours, rainfall, humidity etc., for monitoring the weather condition and pollution. The absolute or relative concentrations of gasses in atmosphere are used to measure basically for monitoring purpose. However, limited studies were dealt with actual gaseous exchanges of fluxes. The concern about gaseous flux measurement and GHGs-exchanges started after 1900s when people got concerned about anthropogenic pollution, environmental sustainability and climate change issues. The GHGs-exchanges in earth-atmosphere systems primarily occur due to change in meteorological variables, soil responses and physiological activities of crops. If we are interested for environmental data inventories, then absolute GHGs concentrations of atmosphere at a specific time is required. However, presently researchers, students, environmentalists, climate scientists are primarily interested for quantitative data on GHGs fluxes and or GHGs-exchanges. They want to know what and how much quantity of a particular GHG is emitting from unit area at unit time.

Flux refers to the amount of any gas emitted per unit area per unit time. The dimension of that is $M A^{-1} T^{-1}$. (Where, M= Mass/ Quantity; A= Area; T= Time). The SI unit of flux is g $m^{-2}s^{-1}$. Therefore, flux is a vector. It has magnitude as well as direction. As for example positive flux of CO_2 from earth surface means, CO_2 is moving out from earth surface to atmosphere. If it moves in opposite direction then the flux would be negative. In that case CO_2 gas may be absorbed by the soil surface from adjacent atmosphere. Similarly, GHGs fluxes of earth surface are either positive or negative. Generally, positive GHGs flux refers to emission of GHGs from earth to atmosphere and vice versa.

In nutshell, the GHGs measurements in agriculture are based on two fundamental principles. First one is the collection of gases from target location (field/experimental site) and subsequent analysis by gas chromatography (GC). The second principle is through sensor based analysis either by infrared gas analyzer (IRGA) or with the help of specific gas analyzer.

In agriculture, continuous and precise measurement of the GHGs emissions from different major cropping systems are necessary for proper quantification and budgeting. Therefore, not only high frequency but also long-term measurements of GHGs fluxes are required. Different technologies with different working principles are available for measuring of GHGs fluxes / emissions from agriculture. The three basic and widely used techniques are 'manual close chamber method', 'automatic chamber method' and 'sensor based systems'. Conventionally, measurement of GHGs emissions from crop field is done with the help of 'manual closed chamber method' for CO_2, CH_4 and N_2O. However, for CO_2 measurement from soil and plant canopy, 'respiration chambers' using infrared gas analyzer (IRGA) is preferred by scientists, students and scholars. High frequency chamber data integrate over small areas could precisely quantify GHGs fluxes. However, it is associated with errors related to perturbations of the natural conditions inside the chamber at the sampling site, changes of microclimate inside chamber, pressure-induced gas flows inhibition resulted concentration build-up in closed chambers. Even though, this methodology is widely followed in agriculture for GHGs flux measurement with site specific modifications. Moreover, methods should be chosen based on the objective of the study and the precision required. The gases to be measured, cost and spatial coverage are other three factors which determine the choice of method for measurement. Further, properties/ processes of soil- plant-atmosphere continuum must be understood clearly for proper interpretation of observed GHGs flux data. Three techniques widely used to measure GHGs fluxes/ emissions in agriculture are described in details in following sections

4.1 Manual close chamber method

Greenhouse gas samples are first collected in the field from soil-plant systems with the help of specifically designed close chamber and then the concentrations of collected gas samples are measured through gas chromatography. The net flux or emission from soil-plant systems then estimated as a change in concentration of desired gas at unit time (gas collection time in the closed chamber) in unit area (area represented by manual close chamber).

4.1.1 Working principle and basic components

Greenhouse gas emission in the soil-plant systems could be estimated by 'manual close chamber method' with the help of a close-chamber by collecting the gas samples frequently from the field and subsequently, by measuring the change in concentration of the target gas. The fluxes of the specific GHG during the study period is estimated with the help of change in concentration of the

GHG per unit area per unit time assuming the linear change in concentration over the period (Hutchison and Mosier 1981, Bhattacharyya et al. 2014, Bhatia et al., 2010, Bhattacharyya, 2016).

The chambers are preferably constructed by multilayered polycarbonate or perspex sheets. However, acrylic sheet or rigid plastic are also used. Based on the objective of the study and crop type dimensions of chambers vary. Even dimensions are changed with crop growth stages, canopy spread and height. As for example, relatively small chambers are used for high yielding rice, wheat, pulses and oil seeds, whereas, long chambers are suitable for maize, barley, sugarcane, and horticultural crops. A typical chamber dimension commonly used for high yielding puddled rice cultivars for collecting gases from panicle initiation (PI) to harvesting stages is 57 cm × 37 cm × 71 cm (length x width x height) (Figure 4.1, 4.2 and 4.3). In majority of the cases rectangular or square chambers are used. However, cylindrical chambers are not uncommon with varying diameter and height. Base plates with sharp base and channel (Figure 4.4) is another basic and important component of this method must be placed permanently in the field before sowing/ transplanting of the crop. Ideally, the base plate position should not be altered during study period to avoid the external disturbance. It is generally made up of aluminium frames/stable plastic pipes with desired dimension, so that chamber can suitably placed on that. It must be inserted to 6-12 cm inside the soil, based on soil type, method and amount of irrigation and type of crop. The base plate-channels should be filled with water before putting the chamber on it to make the base plate-chamber system air-tight.

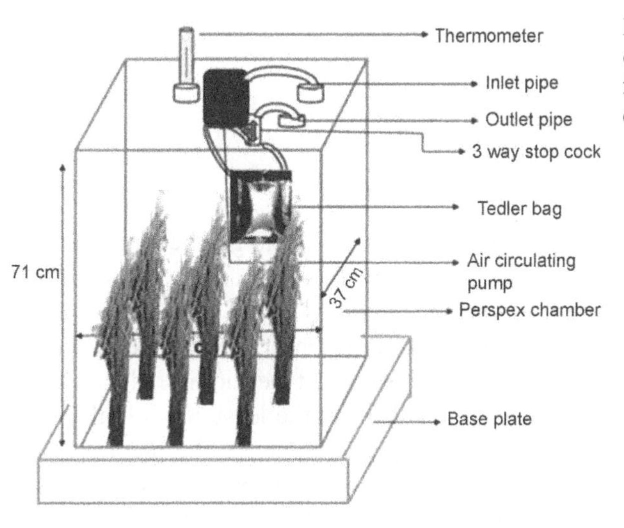

Figure 4.1 Schematic diagram of close chamber fitted with pump for air circulation

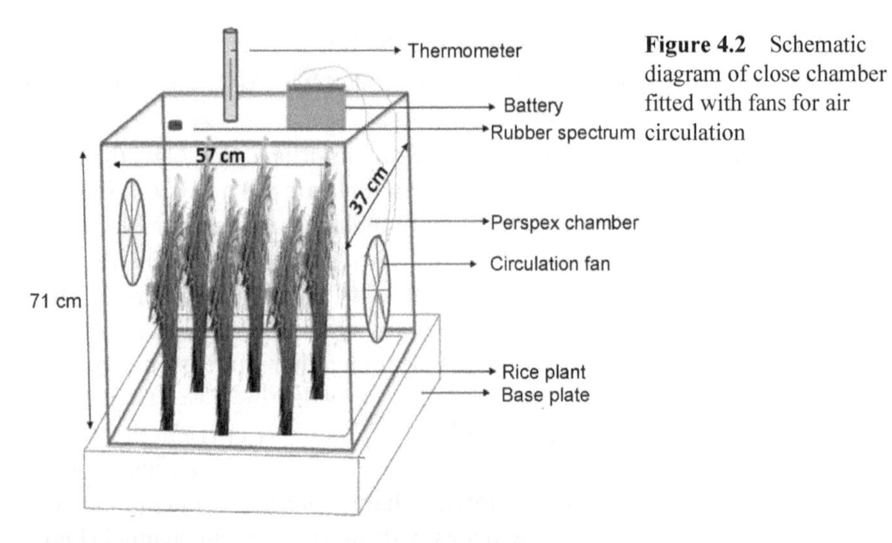

Figure 4.2 Schematic diagram of close chamber fitted with fans for air circulation

Figure 4.3 Photograph of close chamber fitted with pump for air circulation in rice field.

The mixing of air inside the chamber must be ensured to get the representative homogeneous gas samples. Therefore, either fans (two or more at top and bottom) or pump need to be used to homogenize the air inside the chamber during sampling. As the expansion and contraction of the volume of the gas depends on temperature and pressure, those must be monitored inside the chamber during the sampling period. Last but not the least, the initial as well as high frequency periodical sampling is prerequisite for the precise GHGs measurement. The second step is the estimation of concentrations of GHGs of the collected gas samples with the help of gas chromatograph (GC). Immediate analysis in GC is necessary to prevent the diffusion loses of gas concentration.

4.1.2 Components and their specification

1. *Manual closed chamber:* The materials could be used for making the rectangular/ square/ cylindrical static chamber with at least 90% light transparency for maintaining normal photosynthesis of plants, are polycarbonate, multi-layered polycarbonate, perspex, rigid/hard transparent plastic or acrylic sheets. Students or researches can chose any one of those depending upon the local availability and economics. However, for precise measurement, multilayer polycarbonate and perspex chamber are advisable due to their transparency, durability and moulding features. Usually 5-8 mm thick sheets are preferred for close chamber. The dimension and shape of the chamber depends on objective of work, crop type, stages of crop growth, spacing and the method of irrigation followed. As for example, two different dimensions of chambers are generally used in lowland rice-paddy depending on the stage of crop growth for measuring periodic GHGs fluxes during the crop growth season. One is for seedling to panicle initiation (PI) stage (length × width × height: 57 × 37 × 53 cm), and another for panicle initiation to harvesting (length × width × height: 57 × 37 × 71 cm). But it is not fixed. User can modify according to their need; however, he or she has to take care of its exact area and volume during flux calculation. One of such example in 2 meter height chamber for maize and circular shaped chamber for horticultural crop. Even smaller (30 cm height) square shaped chambers are preferable in grassland. Importantly, the shape as well as dimension of the close chamber should be carefully measured and used during calculation of fluxes of GHGs.
2. *Base plate*: The base plate has two basic parts, one in sharp edge-bottom in order to penetrate in soil and another one is broad channel/ grove to hold the chamber. It is generally made of aluminium frame or hard/stable plastics, so that it should not interfere with the ionic-elements in soil, long lasting, non-reactive, relatively cheap and locally available (Figure 4.4). Base plates are generally inserted at least 8-10cm inside the soil. It must be kept undisturbed throughout the cropping season or study period in order to maintain natural soil condition, microclimate and plant performance. It also reduces the spatial sampling errors and helps to overcome the soil heterogeneity influences on GHG production and transmission in soil-plant system. Sampling error and plant performance bias also could be eliminated with fixed base plate technique. During measurement the channels/groves on the base plate must be filled with water before placing the chamber. It is required to make the chamber-base plate system air tight at the time of gas sample collection. It is vital

to curtail down the ambient air interference during gas collection and to prevent leakage.

3. *Thermometer*: The initial and final as well as intermediate temperatures inside the chamber should be monitored during sampling of gases from the field. As the temperatures in field condition widely varies with water management, time of sampling and seasons. The temperature has significant effect on the volume expansion and contraction of gases and subsequently the fluxes of GHGs from the agricultural field. Temperature need to be considered for volume correction in flux calculation. Precise thermometer (°C/°F) can be used for measuring temperature inside the chamber during GHGs collection.
4. *Pump and Fan*: Proper and continuous mixing of air inside the chamber during the sampling period is necessary for collecting representative sample at a particular time. Pump (0.5 litre/ min) either battery or solar power operated or fans (two or more) (Figure 4.4) is used to homogenize the air inside the chamber. Pulse pump is generally used which can be fitted outside the chamber. However, now a days battery operated fans which are fitted two opposite sides (at different heights) at the inside wall of the chamber are preferred.
5. *Scale:* Scales with appropriate range (may be of 0-50 cm and 0-150cm) are required for measurement of plant parameters, depth of the water inside the chamber and outside field.
6. *Tedlar bag*: Some scholars use tedlar bags for collecting gas samples from the field. There are two reasons for that, first one; it helps in getting homogeneous samples and the second one it allows to get sufficient amount of representative samples at a particular time. However, presently gas samples are collected directly from chamber through septum in 50ml syringes fitted with three way valve for convenience and easy handling.
7. Syringe and needles: These are used to collect gas samples from the tedlar bag (Figure 4.4). At the same time syringes are also directly used to collect GHGs samples from the chamber though septum fitted at the top of the chamber. Gas samples are drawn with the help of needle fitted in syringe at 0, 15 and 30 minutes or 0, 10, 20, 40, 60 minutes interval, (based on objective of the study) after putting the chamber on the base plate.
8. *Septum and Stopcock:* Silicon septum is generally fitted at the top of chamber to collect gas samples. Three-way stopcock is used to make the syringe air tight after drawing the gas sample.

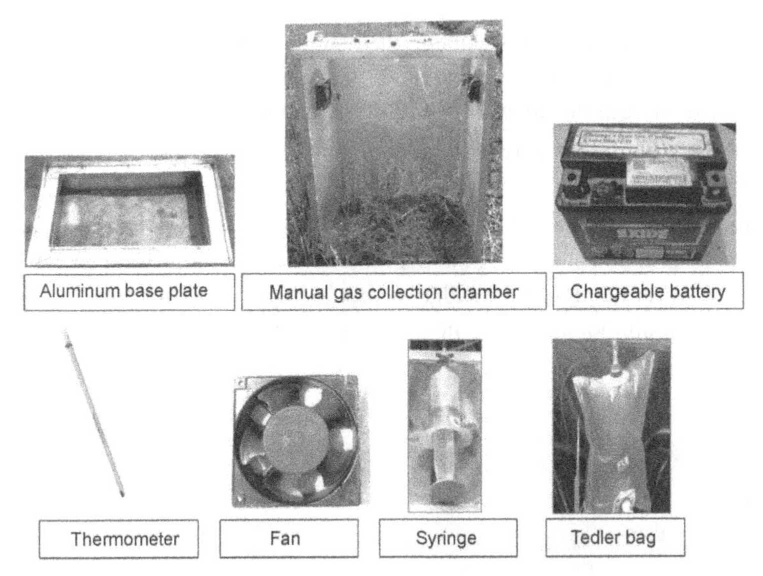

Figure 4.4 (a) Manual gas collection chamber and accessories

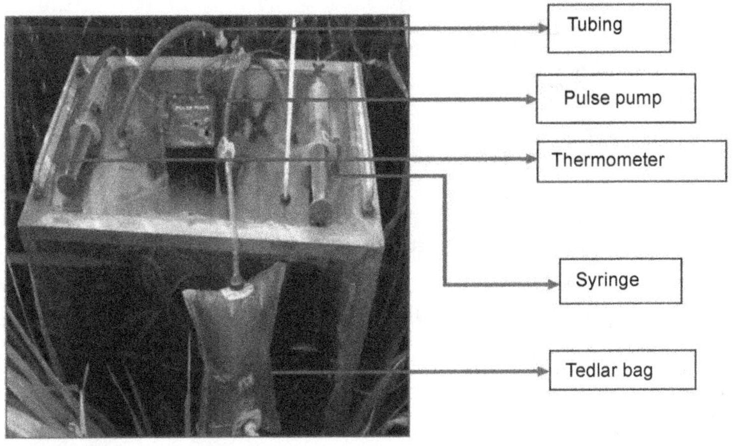

Figure 4.4 (b) A close view of close chamber fitted with syringe, pulse pump, tedlar bag, tubing and thermometer.

4.2.3 Step by step procedure of GHGs collection by manual chamber method

Step 1: Placement of base plates in targeted location (specific treatment with replications), where we want to take measurement prior to sowing/ transplanting.

Step 2: Pouring the water in the groves or channel just before placing the chamber on base plate.

Step 3: Placing the chamber on the base plates.

Step 4: Observe the temperature of the inside the chamber at zero minute. At the same time record the depth of ponding in the field. That is the height of water in the field both inside and outside of the chamber.

Step 5: Collect the gas sample at zero minute with the help of syringe through the septum or from tedlar bag.

Step 6: Keep the chamber in the field for 15 and 30 minutes. And collect the gas samples at 15 and 30 minutes again with the help of syringe. Temperature and ponding water depth of the field also should be noted at that time.

Step 7: The syringe should be air-tied with stopcock immediately after collecting the gas sample (Fig. 4.5).

Figure 4.5 Collection of gas samples from rice field using close chamber method

4.2.4 Points to be remembered during gas sampling

1. Chamber size should be such that minimum representative plants are accommodate there. As for example in case of rice, six hills must be there inside the chamber.
2. Zero minute gas sampling is important, it should be measured precisely. So first gas sample inside the chamber should be taken immediately after placing the chamber on the base plate.
3. Gas samples should be taken in reasonable time intervals (15, 30 minutes or 10, 20, 30, 40, 60 minutes) based on the objective of the study. So that fluctuations of gas concentrations in headspace could be detected precisely. Very short or longer chamber deployment durations may result erroneous output.
4. More number of deployments of chambers is advisable in order to reduce the spatial variability.

5. The temporal variability should be reduced by increasing the sampling frequency.
6. Longer period (30 minutes) of sampling results in better precision however too long (> 2 hour) period may result sampling artefacts.
7. To avoid the diurnal variation, the gas samples should be collected twice in a day preferably at 9-11 AM in morning and 3-5 PM in the evening.

4.2.5 Analysis of GHGs by Gas Chromatography (GC)

Methane, nitrous oxide and carbon dioxide concentrations of the gas samples collected through "manual chamber method' could be precisely analyzed by gas chromatography (GC) equipped with flame ionization detector (FID), electron capture detector (ECD) and flame ionization detector (FID) fitted with a methanizer, respectively.

The flame ionization detector (FID) is generally used for detection of hydrocarbons that generate ions when heated with H_2-air flame. The basic principles are, at first gas enters in to the hydrogen jet, *via* filter (millipore) with a carrier gas then get ionized. After that the free electrons are entered in to the flame at the tip of the jet. As the generated electrons are drawn towards the collector there would be a continuous flow of current. The voltage drop due to flow of current across the resistor is being amplified and displayed at logger/ recorder. The entire FID system along with jet, resistor, amplifier, and recorder is housed in an oven where a specific temperature is maintained in order to prevent the condensation of water vapour around the detector. A Porapak-N or Porapak-Q column with ~3-m-long nickel or stainless steel made having ~3.18-3.20 mm outside diameter is usually used. The condition or method generally practiced in GC-FID detection of methane are column temperature around 70°C; helium, nitrogen or argon as carrier gas with a flow rate of ~ 20-30 $cm^3\ min^{-1}$; detector temperature is around 250°C; Hydrogen (H_2) flow rate of ~30-40 $ml\ min^{-1}$ for ignition (Figure 4.6). However, method should be adjusted/ optimized according to GC's internal configuration, column length, column material used, and the objective of our study. The sampling valve could be adjusted, electronically, pneumatically or manually. The GC- microprocessor based software basically used for chromatogram plot analysis and peak area determination. The calibration could be performed with the help of pure (99.999%) methane standards like, 1, 2, 3, 5, 8, 10 and 20 ppm based on the our expected unknown sample ranges. After calibration the gas samples collected through 'manual gas chamber' containing methane are introduced (gas sample loop of 1 or 2 cm^3) to GC by a syringe to the injection port.

Figure 4.6 Gas Chromatography with FID detector for analyzing the methane concentration.

The electron capture detector (ECD) is used to measure the concentrations of N_2O in the collected gas samples in GC. Generally, a Porapak-Q column of 6 feet long with 1/8 inch outer diameter and SS (stainless steel) column (80/100 mesh) is used for separation. Basically the detector has two electrodes one of which is treated with radioactive titanium or scandium that emits beta rays and other one is positively charged polarized electrode. The high-energy electrons generated from radioactively treated electrode bombard with the carrier gas (nitrogen or argon) and produce a large numbers of thermal electrons (low energy secondary electrons). Then those thermal electrons are collected by the positively polarized other electrode of the detector and there is steady state of current flow. This current flow between the electrodes is reduced/ changed when an electrophilic gas/substance passing through the space between the electrodes. It captures some of those electrons and generates an electrical pulse of the chromatogram peak. So, the point is to be remembered that only the electrophilic substances could be detected by ECD. The condition or method generally practiced in GC-ECD detection of nitrous oxide are column temperature around 60°C; nitrogen as carrier gas with a flow rate of ~ 15-18 ml min^{-1}; injector temperature ~ 200°C; detector temperature is around 340-350°C. The GC-ECD system must be calibrated before as well as after the measurements by using ~ 100-110 ppb (parts per billion) pure N_2O in N_2 as the primary standard and ~ 300-320 and 390-400 ppb N_2O in N_2 as the secondary standards (Bhattacharyya et al., 2016). Instrument-specific advanced software may be used for plotting the chromatogram and peak area.

The 'GC- FID- methanizer' system is generally used to measure the CO_2 concentrations of the collected gas samples through 'manual close chamber method'. The detail configuration of FID with condition/ method is described in previous section. The additional attachment of the system is methanizer that consists of a stainless steel (SS) tube (~ 6" × 1/8") mounted on the edge of the heated valve-oven. The SS tube is generally packed with catalyst powder made up with activated zinc/nickel/platinum-lead. The hydrogen gas mixed with column effluent and entered in to the methanizer at a flow rate of ~ 20 ml m^{-1}. The column temperature is maintained around 380°C. Primarily, the CO and CO_2 in gas samples are reduced and converted to CH_4 in methanizer and after that CH_4 concentration is detected by the FID in GC. The methanizer is unaffected by hydrocarbons like propane, ethane and methane. The tubes of methanizer could be poisoned by sulphur gas.

4.2.6 Advantages and disadvantages of 'manual close chamber method'

Advantages

1. The chambers and accessories are simple, economic and easy to make with locally available material.
2. The size and dimension of the chamber is flexible and could be modified according to the need of the study/work.
3. Treatment differences could be captured even in small plots and moreover, very small changes in fluxes can be detected.
4. It is a standalone system with a small battery could be solar operated and no extra power supply is required.
5. It is easy to handle so no extra skill or specific training is required to collect the samples.
6. Little disturbance to the crop during the sampling.

Disadvantages

1. End point inhibition on fluxes may be happened due to accumulation of gases during sampling time. So normal emission rate might be distorted.
2. Enhancement of temperature and or pressure inside the chamber during sampling time may lead to underestimate the fluxes.
3. It would give only point data for a specific time, so diurnal variation rarely capture unless high frequency sampling is done throughout the day which is laborious and difficult to execute in field condition.

4.3 Automatic chamber method

The basic working principle of 'manual chamber method' and 'automatic chamber method' is mostly similar. The primary difference is that continuous collection and analysis of GHGs fluxes are performed in 'automatic chamber method'. There is a provision of automatic opening-closing of the lid of the gas chamber but chambers are placed in permanent position in the field above pre-installed base plates (Figure 4.7). The gas samples are collected in fixed time intervals with the help of a suction pump and subsequently analysed in gas chromatography.

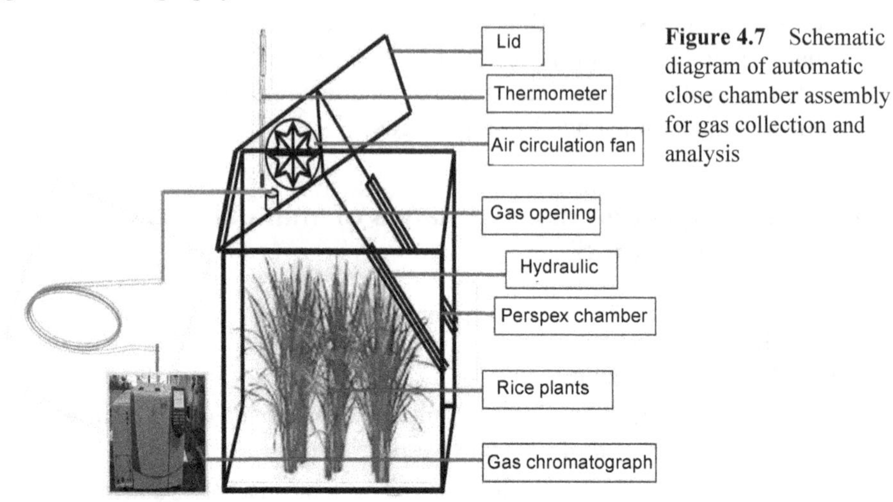

Figure 4.7 Schematic diagram of automatic close chamber assembly for gas collection and analysis

4.3.1 Working principle and basic components

In 'automatic chamber method,' chambers are not portable like 'manual chamber method'. Here initially base plates are installed in the field with the chambers in the targeted location. The base plates and chamber are attached with each other unlike 'manual chamber'. Therefore, the requirement of number of base plates and chambers are equal in this method. The chamber material are same as of 'manual chamber' i.e. either polycarbonate or acrylic or Perspex sheet. The automated opening-closing lid is the basic difference of this chamber-method to 'manual chamber method'. The lid is made up of same material as the chamber. At least two fans are fitted at opposite sides of the chamber for circulating the air inside the chamber for proper mixing of gases. The opening and closing of lids in the chamber and operating fan just 4-5 minute before sampling are fully automatic and controlled by computer software. Like 'manual chamber' there is no need of tedlar bag, syringe and septum opening to collect the gas samples. In 'automatic chamber' the suction pipe of GHGs analyzer is fitted at the ceiling of the hood. The gas samples are collected at specific intervals by suction pumps and transferred to the input port of GC for real time GHGs analysis. The sampling intervals could be chosen through computer software -programmes by the students/ researches or users according to their objective of study. The detector used and method followed to analyse methane, nitrous oxide and carbon dioxide are same as discussed in earlier section (section 4.2) in 'manual chamber method' (Mukherjee and Sarkar, 2008; Mukherjee et al., 2009; Bhattacharyya et al., 2012, 2016).

4.3.2 Step by step procedure of GHGs collection by 'automatic chamber method'

Step 1: Placement of 'base plate-automatic chamber' assembly in targeted location (specific treatment with replications), where we want to take measurement prior to sowing/ transplanting.

Step 2: Checking the leakage of movable lid and joints of base plate-chamber assembly.

Step 3: Setting the method in computer software for gas sample collections cycle/ intervals and put the automatic system on.

Step 4: The GC also must be on with control unit for real time analysis.

Step 5: Recording the inside temperature of the chamber at each sampling cycle including at zero minute. At the same time record the depth of ponding in the field. That is the height of water in the field both inside and outside of the chamber.

Step 6: Ponding depth as well as the height of water in the field both inside and outside of the chamber should be recorded manually also during sampling period.

Step 7: Gas sampling could be done throughout the day at specific intervals depending on the objective of the study and user requirement.

4.3.3 Points to be remembered during gas sampling

1. There should not be any leakage in the junction of lid and body of the chamber and the base plate-chamber joints. Both the portion must be made air tight.
2. The suction pump-pipe line should be obstruction free.
3. There should be synchronization between the lid-opening of the chamber and gas suction (gas sampling through suction pump), so that minimum time lag would be there at specific time of sampling.
4. The collected gas from the chamber should be filtered before entering in to GC input port.
5. The voltage fluctuation and inter-phase functioning should be monitored carefully during sampling.

4.3.4 Advantages and disadvantages of 'automatic chamber method'

Advantages

1. It is expected to give more precise results than that of 'manual chamber method'.
2. The diurnal variations in fluxes of GHGs can be effectively captured by this method.
3. It provides real time GHG flux data. That is, there is no significant lag between gas collection and analysis by GC. So, there would less chances of leaking from collected gas samples in syringe.
4. This chamber method for GHGs fluxes are often used field trials having multiple small plots.

Disadvantages

1. This method is relatively costly than manual chamber method.
2. Continuous undisrupted power supply is required for this analysis.
3. Automatic chamber method some time suffers from underestimation/ overestimation of fluxes due to chamber effects and soil moisture conditions during rainfall.

4.4 Infrared gas analyzer method

The carbon dioxide fluxes are measured through Infrared gas analyser (IRGA) method. It is widely used portable system, primarily used for soil respiration and photosynthetic activities of plants. However, till now its application is restricted to estimation of CO_2 fluxes and not effectively used for methane and nitrous oxide measurement.

4.4.1 Working principle and basic components

The CO_2 flux refers to the change in concentration of CO_2 per unit area per unit time. In this method it is measured by placing an opaque chamber (specifically known as soil respiratory chamber) on the soil with or without any canopy for measuring the change in CO_2 concentration at a definite time interval. In general the soil respiration system consists of a soil respiration chamber (SRC) for collection of gases and an analyser to measure the gas concentration. The chamber volume is known and could be modified according to user need. The continuously sampled air from the chamber having a closed circuit is analysed by the IRGA. The system must be calibrated frequently before the field measurement with a known CO_2-concentration environment. The CO_2

fluxes are calculated pre-determined programme in data logger which is an integral part of the system and also stored in the logger.

In this method, the concentrations of the CO_2 of the specifically designed chamber are measured through IRGA. The sealed opaque chamber, suction pipe for transferring gases to detector, filters, data logger and battery (power supply unit) are the basic components of this system. It is portable, so very effectively used in different treatments in the field condition. In infrared gas analyzer, actually increase in CO_2 concentration in enclosed chamber over a specified time is measured. The CO_2 effluxes measured by IRGA are the function of chamber area and volume, and time intervals of gas collection. Therefore, the gas flow rate to IRGA, enclosure time and purging mechanism are also important driver of CO_2 flux estimation by IRGA-method. The IRGA is sensitive to moisture and temperature, so necessary precautions and corrections must be employed during CO_2 flux estimation. The chamber type and time of enclosure (enclosed time of chamber for gas collection) should be customized by the user based on the objective of the study, soil and crop type (Figure 4.8).

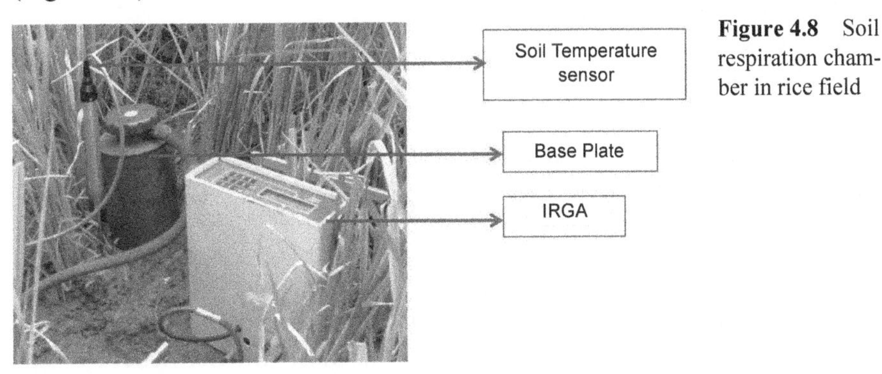

Figure 4.8 Soil respiration chamber in rice field

4.4.2 Step by step procedure of gas analysis by IRGA-method

Step 1 Placing the base plates in target location (plots where we want to take measurement) well before the experiment started or planting.

Step 2 The portable chamber would be placed on base plates when we want take the observations.

Step 3 The IRGA- equipment's sensor must be on along with the data logger after placing the portable soil respiratory chamber or gas chamber on the field.

Step 4 The data of CO_2 flux automatically stored in data logger with specific unit as customized by the user. The data could be stored in data logger and retrieved after words.

4.4.3 The points to be remembered during analysis

1. The air within the chamber should be mixed properly in order to ensure representative gas sampling without generating pressure differences.
2. The system must be leaked proof to ensure the linear rate of increase of CO_2 over enclosure time.
3. The moisture free gases should be introduced in IRGA through proper filtering.
4. The chamber size and dimension are flexible; so may be customized with crop type, growth stage and target area. As for example, for soil respiration study chamber height is only 20cm while for canopy CO_2 flux measurement of wheat it is around 75-90cm.
5. Base plates of desired dimension should be used to protect the chamber and also to get undisturbed gas samples. Base plate should be placed well before sampling similar to that of "manual chamber method' (described in section 4.1).
6. Calibration of the system is the prerequisite and must be done for each separate field experiment.
7. IRGA system should be protected from water and rain as it is very sensitive to moisture.

4.5 Photo acoustic spectroscopy

4.5.1 Working principle and basic components

The working principle of photo acoustic spectroscopy is based on Bell's theory. In this method acoustic detection is done on the effect of observed radiation (electromagnetic) on substrate. It is useful in *in-situ* measurement and provides precise result for trace gases analysis. Another benefit of this method is that it is non-destructive. However, some modifications were made with the basic components to make it more users friendly and sensitive. The modifications are listed below:

1. Now, in 'laser photo acoustic spectroscopy' laser is used instead of sun to illuminate the target object to get proportional sound generation with light intensity.
2. Sensitive micro-phase with amplifier (lock in amplifier) has been used in place of 'ear' in previous instrument.
3. Sound signal is now amplified by enclosing the sample with a cylindrical chamber. The modulation frequency of the sample is also turned to get an acoustic resonance.

4.5.2 Basic components

The CO_2 laser as radiation source: The tuneable CO_2 laser (9-11 Am) is used for molecular finger print (electromagnetic radiation source). A high power (> 10 Watts) gas laser is used which could be able to give proportional signal intensity with emitted light energy with a sensitivity range to ppm. Nevertheless, these lasers have two disadvantages, they are large and expensive and thus its use is limited to laboratories.

Quantum cascade laser (QCL): The quantum cascade laser (QCL) is used as a source of radiation. This source has compact size, with continuous emission and high spectral resolution. It can also be operated in room temperature (20- 25°C).

Detector for GHGs: Photo thermal detector is used for CO_2 laser photo acoustic spectroscopy. This photo acoustic spectroscopy could also be used to detect volatile organic compounds (VOCs) emissions, such as ethylene (Moraes et al., 2003; Fejer et al., 1992, Zhang et al., 1993). Methane is detected by employing two quantum cascade lasers in the range of 7.7-7.9 Am and can generate a power of 5.6 mW.

4.5.3 Step by step procedure for measurement

Step 1: Gas sample is collected and sealed into photo acoustic chamber.

Step 2: The samples are then irradiated with IR light with a frequency in which resonance with the frequency of standard GHG molecule.

Step 3: By absorbing some portion of IR radiation the gas molecules gain heat energy and subsequently the temperature and pressure of the sample gas is also increased.

Step 4: The pressure variation of the chamber creates acoustic wave, when IR radiation is modulated with definite frequency.

Step 5: Then the acoustic wave is converted in to electrical signal by microphone and detected.

4.5.4 Advantages and disadvantages

Advantages

1. It is a portable system, so treatment differences could be captured at field level.
2. All important GHGs could be measured by this method with high precission.

3. Unlike IRGA-CO_2 system, it could detect the CH_4 flux in field at ppm level.

Disadvantages

1. The CO_2 laser radiation sources are large and expensive.
2. Quantum cascade laser is relatively less expensive but not so precise.
3. Different set of power and light are required for different trace gases.
4. High skilled and expertise are required for running the instrument.

4.6. Eddy covariance system for GHGs measurement

4.6.1 Working principle

The eddy refers to the small three dimensional pocket of wind that has both magnitude and direction. The eddy covariance method is actually the covariance between vertical component of wind speed and concentration of desired gas at a particular moment. It is a micrometeorological technique generally uses to estimate the turbulent fluxes of GHGs (CO_2, CH_4, etc.), water vapour, heat and organic volatile compounds in the boundary layer of atmosphere. This technique or method is also popularly known as 'Eddy Flux technique'. It gives direct measurement of turbulent flux of a gas (scalar, e.g. CO_2/ CH_4 / H_2O vapour) across the horizontal wind streamline near or above plant canopy (Baldocchi et al., 2000). In this method high frequency (10, 20, 40 Hz) time series data of wind speed and gaseous concentrations are used for flux calculation. The positive 'eddy flux' represents the emission (outgoing) of GHGs / gases from the system at sensor height, whereas, the negative eddy flux indicates the uptake (absorbed) of GHGs/gases in the system. It is an advanced, highly precise, stationary, real time system widely used throughout the world.

4.6.2 Basic components of Eddy Covariance system

The basic unit of eddy covariance system includes three dimensional sonic anemometer, carbon dioxide and H_2O gas analyser, data logger, power supply unit, either flux tower or tripod for sensors holding and data processing soft wars.

(i) Three dimensional sonic anemometer

Three dimensional (horizontal, vertical, and longitudinal) wind speeds is measured by sonic anemometer (Fig. 4.9a) with high frequency (10 Hz), i.e., millisecond data.

Figure 4.9a Three dimensional sonic anemometer

(ii) Carbon dioxide and H_2O gas analyser

High frequency and sensitive infrared gas analyser (IRGA) is used for measuring the concentration of CO_2 and H_2O vapour. It is a sensor based technique. Calibration of the analyser should be done against the standard, i.e., pure CO_2 (99.99%) for carbon dioxide (Fig. 4.9b).

Figure 4.9b Carbon dioxide and water vapour analyser sensor

(iii) Data logger

Data logger is an electronic device used to collect, store and retrieve large amount of time series data measured by sensors (Sonic anemometer, IRGA). Actually it is an analyzer-interface unit to collect and store data by the Ethernet (Figure 4.9c).

Figure 4.9c Data logger of eddy covariance system

(iv) Power supply unit

Eddy covariance system is a standalone stationary system. It gets power from battery which could be recharged through appropriate solar panel. Generally, a well designed photo voltaic power system is used to supply continuous power to the system (Fig. 4.9d).

Figure 4.9d Solar panel and chargeable battery

(v) Flux tower and tripod

The eddy covariance system could be installed either on flux tower (2 m of each stretch/ compartment) at around 10, 20 or 30 meter height based on the canopy height or in tripods (2-3 meter height). The flux tower primarily installed in forest ecosystem and horticultural or agro-forestry system. While, tripods system is used in agricultural crops and pastures. There are well networked flux system exist in Europe, Asia and America named as 'Euroflux' (CARBOEUROPE), 'Asiaflux' and 'Ameriflux', respectively. And the data bases are popularly known as 'FLUXNET' database.

(iv) Data processing software

Eddy covariance systems generate huge volume of time-series data (high frequency data). It needs user friendly, flexible software for processing the raw data considering the relevant corrections. The software should have platform to import different time series data format, allow configuring sampling rate and configuring auxiliary sensors inputs. Currently, many types of software are available for processing high frequency time series data. These are TSA, TK2, EdiRe, EC pack, EddySoft, EdiPro etc. Most of the software used more or less similar basic empirical model/equations/processes/ approaches for processing the data with corrections. Therefore, output flux data generates not varied significantly. Only differences exist among the software, are in processing speed, number of corrections considers, type of input data used and numbers of estimated outputs.

4.6.3 Sensor used in Eddy covariance system

Apart from CO_2/H_2O analyzer (IRGA) and sonic anemometer the other sensors could be intergraded in eddy covariance system (Figure 4.10). Few other sensors are:

i.	Methane analyser	ii.	Air temperature and relative humidity sensors
iii.	Net radiation sensor	iv.	Canopy temperature sensor.
v.	Four component radiation sensor	vi.	PAR sensor
vii.	Soil heat flux plate	viii.	Snow depth sensor
ix.	Time domain refractometer	x.	Tree-Bale temperature
xi.	Barometer processer	xii.	Surface temperature sensor

Figure 4.10 Typical open path Eddy Covariance System in rice field

4.6.4 Eddy covariance data processing: Assumption calculation, corrections and partitioning

Large amount of time-series data is generated by eddy covariance system. Initially these data are saved in a specific data format (typical data format, e.g TOB 3 file format., etc) based on data logger capabilities. After that, those data is retrieved and processed through flux processing software (EdiRe, Edipro etc). Most of the time, software is taken care of correction, calculation and quality checks of time-series data by following definite algorithm (with empirical or process based model/routine/sub-routine) through software. However, processing software could be customized according to the need of the users and the objectives of the studies.

Assumption of Eddy covariance data processing

The major assumptions of EC data processing are,

1. Flux should be turbulent and majority of net vertical transfer are done through eddies.
2. Measurement must be accomplished within atmospheric boundary layer.
3. Fetch area should be horizontal and uniform.
4. Density fluctuations, flow convergence – divergence within fetch should be negligible.
5. Fetch area should be large enough in upwind direction at the area of interest.
6. EC system should detect minor changes of scalars and wind speed at very high frequency (5 to 40 Hz)

Calculation of flux

The flux is calculated with the basic principle of estimating covariance between vertical wind speed and scalar (e.g. $CO_{2,}$ CH_4, etc.) of time-series data. Therefore, the net ecosystem CO_2 exchange (NEE) is calculated to sum-up the eddy CO_2 flux (F_C) and CO_2 storage (F_S). In most of agriculture crops F_S is negligibly as canopy height are relatively less than forest and other horticulture crops.

The mean vertical flux density of CO_2 is calculated by the formula,

$$F_C = \rho a * w^{/} c^{/}$$

Where, F_C = mean CO_2 flux; w = 30 minutes' covariance between vertical fluctuation, **c**$^{/}$ is the CO_2 mixing ratio; ρa is the air density ratio; 'Over bars' denotes 'time average'; and primes refers to fluctuations from average value. A positive value of flux indicates emission of CO_2/GHGs from soil-crop system to atmosphere and negative value represent CO_2 absorption/ uptake by the soil-crop system from atmosphere. Generally, 30 minutes or 1 hr average of NEE is utilized for budgeting of annual or seasonal net exchanges.

Correction of EC data for quality check

Generally various potential flux estimation errors can be introduced in time series data due to instrumental limitation, assumptions, terrain features and distortion of physical environment at measurement sites. The primary errors are generated through time response, sensor separation, tube attenuation, high and low pass filtering, scalar path averaging, mismatch in sensor response, time delay of sensor, spikes, improper levelling of instrument or sensors, density fluctuation of scalar due to temperate pressure fluctuation, lower turbulence, low wind speed etc. However, to minimize the errors a number of correction must be done on time-series data before further processing and analyzing the flux. Those corrections are the pre-requisite of eddy flux estimation and without performing those corrections, the EC flux data are not accepted by scientific community. The five major correction namely 'delay', 'despiking', 'planner fit/ co-ordinate rotation', 'detreding' and ' WPL term' are briefly discussed in coming sections.

The error 'delay' or 'time lag' is generally caused by differential electronic signal between two sensors, namely three axis sonic anemometer and CO_2 / CH_4 analyser. It could also caused by spatial separation of sensor and tube length of 'closed-path EC system'. The 'time lag' can be estimated for each averaging interval by performing the cross correlation between scalar of interest and vertical wind speed. The 'time lag' is selected for a specific time period that is having highest cross-correlation coefficient. After that 'time lag' is introduced to correct the error, 'delay'. Spikes in time-series high frequency data could be caused by electronic (instrumental) malfunctioning or by some perturbation of the measurement volume. Algorithms that detect spikes and rectify those with statistical logic are called 'despiking'. However, spikes could be either removed or flagged for 'gap-filled'. Another important correction of EC time-series data is 'planar fit/ coordinate rotation'. The correction must be introduced because it is very difficult to keep sonic-anemometer or other sensor perfectly level on surface and also most of earth-surface is not exactly flat. So, in order to make it streamline coordinate system, we have to do parameterizations that tend to minimize the effect of sloping terrain. So that, the results is comparable to the measurements that are taken over a flat surface. Primarily, the 'coordinate ration' correction is introduced to eliminate errors resulting from imperfectly mounted sonic anemometer. It results in high pass filtering of the scalar-covariance. On the other hand the 'planner fit' correction is essential for uneven terrain (sloppy land) (Wilczak et al. 2001).

The 'detrending' is the processes by which non-fluctuating part of time-series data that do not represent the true turbulence are corrected. This error may be generated either from diurnal changes of scalar concentration, wind speed, or measurement effects due to sudden instrumental drifts. This correction is must as low frequency eddies with period longer than that of averaging period are excluded from the estimated flux. So, this high pass filtering is unavoidable and must be corrected before calculation of eddy flux. The 'detrending' could be done by three techniques like, 'block-averaging', 'linear-detrending' and 'non-linear filtering'. The 'WPL term/correction (Webb-Pearman-Leuning term) is introduced to compensate volume fluctuation of scalars ((CO_2/H_2O/CH_4) and other trace gases) originated due to fluctuation of atmospheric temperatures, pressure and vapour pressure deficit (Webb et al. 1980). This correction/ term are widely used in EC flux data processing. This correction is necessary, because fluctuation of temperature and pressure causes changes in the volume of gases (to be measured) that are not associated with the flux data of trace gases we want to actually estimate.

The EC flux should be estimated only after doing all the necessary correction on time-series data. The flowing flow character (Figure 4.11) is described the EC data processing in a schematic way.

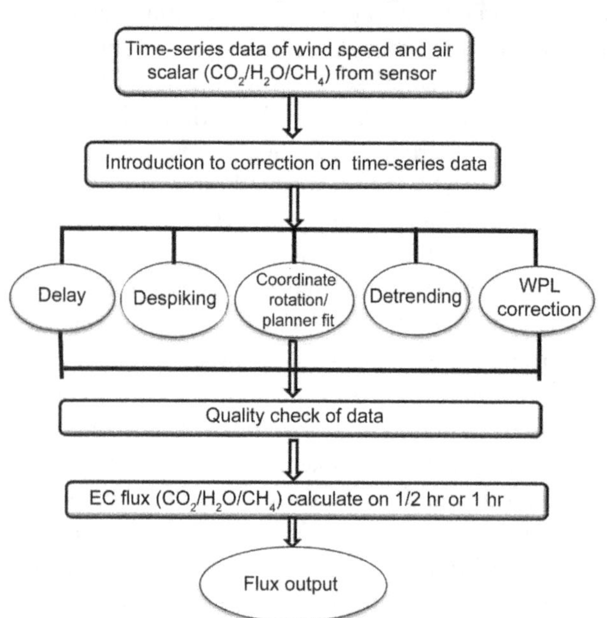

Figure 4.11 Schematic flow diagram of eddy data processing

Partitioning and gap filling of EC flux data

The NEE could be partitioned to gross primary production (GPP) and ecosystem respiration (RE). The GPP refers to photosynthetic assimilation by the crop and RE represent the respiratory fluxes from microbes/ soil or soil fauna (autotrophs and heterotrophs) (Bhattacharyya, et al 2013, 2016). The NEE partitioning could be done by 'Rectangular Hyperbola' method (Ruimy et al., 1995) or by considering night time NEE as RE and extrapolating the relation of air temperature and RE on consecutive day time.

The gap filling of missing or discarded data generally done by following the approaches like 'look-up' table (Flage et al., 2001), 'previous-table-mean' or by model based extrapolation (Logic-based extrapolation).

4.6.5 Advantages and disadvantages of Eddy covariance GHG flux measurement

Advantages

1. It provides high frequency precise data.
2. Non destructive method of measurement; rarely disturbs the natural system.
3. Provide real time *in-situ* data.
4. Reliable and verifiable; It is an automated-data measurement technique.
5. GHGs / carbon foot print analysis could be done very effectively with EC flux data.
6. It could be widely used in forestry, agriculture, horticulture and different land uses with high precision and could provide reliable data inventories in national and international scale.

Disadvantages

1. It is relatively costly instrument.
2. It requires high frequency data measurement capability and un-interrupted power supply.
3. Expertise and definite skilled personnel is essential for data handling and processing.
4. The system is stationary and only provides real time data is a single eco-system at a time.
5. Correction of time-series data is essential; otherwise validity of data would be drastically reduced.

4.7 Equations to be remembered

Calculation of methane flux

$$CH_4 \text{ flux} = \frac{(\Delta x \times EBV_{(STP)} \times 16 \times 10^3 \times 60)}{(10^6 \times 22400 \times T \times A)} \quad (4.1)$$

Where, ΔX = Difference in flux value between 30/15 min and 0 min (converted to ppm based on the standard CH_4), EBV $_{(STP)}$ = Effective box volume at standard temperature and pressure, T = Flux time in min (15 or 30), A = Crop (e.g. rice) area occupied by the box in m^2 (length × breadth).

The EBV $_{(STP)}$ is calculated using the following equation:

$$\frac{P_1V_1}{T_1} = \frac{P_2V_2}{T_2} \quad (4.2)$$

Where, P_1 = Barometric pressure at the time of sampling in mm Hg, V_1 = EBV (Effective box volume), T_1 = 273°K + temperature inside the box at the time of sampling in °C, P_2 = Standard barometric pressure (760) in mm Hg, V_2 = EBV $_{(STP)}$, T_2 = 273°K

The Effective box volume (EBV) is calculated using the following equation:

$$EBV = Box[(H - h) \times L \times B] - V \quad (4.3)$$

Where, H = Box height (cm), h = Height of the water level in the groove of aluminium channel (cm), L = Box length (cm), B = Box breadth (cm), V = Crop (e.g. rice) biomass volume (mL) inside the box (above ground biomass only).

Calculation of N_2O flux

$$N_2O \text{ flux} = \frac{(\Delta x \times EBV_{(STP)} \times 44 \times 10^3 \times 60)}{(10^6 \times 22400 \times T \times A)} \quad (4.4)$$

Where, ΔX = Difference in flux value between 30/15 min and 0 min (converted to ppm based on the standard N_2O values), EBV $_{(STP)}$ = Effective box volume at standard temperature and pressure, T = Flux time in min (15 or 30), A = Crop area occupied by the box in m^2 (length × breadth).

Rest of calculation of EBV $_{(STP)}$ and subsequent portion was same as mentioned in previous section in methane flux calculation.

Measurement of CO_2 flux

The CO_2 concentration in samples can be analyzed using gas chromatograph equipped with thermal conductivity detector (TCD) having 3 m long and 0.3 cm internal diameter HayeSep D column. Helium is used as a carrier gas at a flow rate of 25 cm^3 min^{-1}. Oven and detector temperatures are 50 and

150°C, respectively. Standard CO_2 samples (350, 500 and 700 ppm) are be used for GC calibration.

Flux of gases (F, g CO_2-C m^{-2} day^{-1}) can be computed as:

$$F = (\Delta g/\Delta t)\,(V/A)k \tag{4.5}$$

Where $\Delta g/\Delta t$ is the linear change in CO_2 concentration inside the chamber (g CO_2-C m^{-3} min^{-1}); V is the chamber volume (m^3); A is the surface area of the chamber (m^2) and k is the time conversion factor (1440 min day^{-1}). Chamber gas concentration can be converted from molar mixing ratio (ppm) determined by GC analysis to mass per volume by assuming ideal gas relations. Hourly CO_2 fluxes are calculated from the time vs. concentration data using linear regression.

Calculation of CO_2 flux

A quadratic equation is fitted to the relationship between the increasing CO_2 concentration and elapsed time. The flux of CO_2 per unit area and per unit time is measured using the following equation.

$$R = \frac{(C_n - C_o)}{T_n} \times \frac{V}{A} \tag{4.6}$$

Where, R is the soil respiration rate (flux of CO_2 per unit area per unit time), C_o is the CO_2 concentration at T=0 and C_n is the concentration at a time T_n, A is the area of soil exposed and V is the total volume of the chamber.

Calculation of GWP, CEE and CER

Intergovernmental Panel on Climate Change (IPCC) factors for calculating the combined Global Warming Potential (GWP) for 100 years. The GWP can be calculated by the following equation.

$$\text{GWP} = (25 \times CH_4 + CO_2 + 298 \times N_2O)\ \text{kg}CO_2\ \text{equivalent ha}^{-1} \tag{4.7}$$

Carbon equivalent emissions (CEE) and carbon efficiency ratios (CERs) are calculated by using the following equations (Bhattacharyya et al., 2012, 2016):

$$CEE = \frac{[GWP \times 12]}{44} \tag{4.8}$$

$$CER = \frac{\left[\begin{array}{c}\text{Grain yield (in terms of carbon: carbon content in grain)} \\ \text{of the crop/rice}\end{array}\right]}{CEE} \tag{4.9}$$

4.8 Key messages

1. All the techniques for GHGs fluxes measurement, even the most advanced one also have some limitations.
2. The best approach of GHGs emission monitoring and quantification is the proper integration of the best available methods based on the client demand, and objective of the study.
3. The manual chamber method is most economic and widely used in agricultural system.
4. The eddy covariance system is the most precise real time measurement of GHGs emission at present day.
5. The up scaling of GHGs flux data is an essential requirement, not only for more accurate regional and national inventories, but also for development of site-specific mitigation practices

4.9 Suggested questions

1. Why GHGs measurement should be done on earth surface?
2. What are the major techniques followed to measure the GHGs fluxes in agriculture?
3. Write the principle and step by step procedure to measure the GHGs fluxes by manual closed chamber method.
4. Write the advantages and disadvantages of manual as well as automatic closed chamber techniques of GHGs flux measurement.
5. What is eddy and how the eddy covariance technique is used for measuring the GHGs flux?
6. How the CH_4 and N_2O concentration could be analysed through Gas Chromatography? Explain with calculation and examples.
7. How the soil CO_2 fluxes could be measured by IRGA?
8. Write down the different usage of eddy covariance in agriculture and sensors used in the system.
9. What is the working principle of Photoacoustic spectroscopy? What are the advantages and limitations of this technique?
10. Write major advantages and disadvantages of eddy covariance technique.
11. Write five major corrections done on eddy time-series data for processing.

4.10 Solved problems

1. If the 0 and 30 minutes methane (CH_4) concentration reading in gas chromatograph is 1.869 and 3.158 ppm then what is the flux of methane. The volume of manual chamber used in collection of gas samples from field was 124179 m^3 [71 × 53 × 33 (height × length × breadth)], the final and initial temperature was 32 & 35 °C, there was 2.5 cm standing water in the field and 72 mL is the shoot biomass volume for six rice hills inside the chamber and assume that the pressure inside the chamber was maintained as STP (760 mm) during the collection of gas.

Ans. We know that Flux= Change in concentration of gas / unit area/ unit time.

Step-1: Change in methane concentration over 30 minute = ΔX_{CH4} (30m reading – 0 minute reading) = 3.158 ppm – 1.869 ppm = 1.289 ppm

Step-2: Volume of gas chamber covered= [(H-h) × L × B] = 119806.5 cm^3

Where, [H (height of chamber) = 71 cm, h (height of standing water) = 2.5 cm, L (length of the chamber) = 53 cm and B (breadth of the chamber) = 33 cm]

Effective box/ chamber volume (EBV) without temperature correction (V_1) = [(H – h) × L × B] – V = (119806.5 – 72) cm^3 = 119734.5 cm^3

Where, Volume covered by six rice hills inside the chamber (V) = 72 cm^3 (ml)

Corrected chamber volume (CCV) (temperature correction) (V_2)

$= (T_2 \times V_1)/T_1$

$= [(273x\ 119734.5)/308]$

$= 106128\ cm^3$

[Where, T_1 = 35°C + 273°K = 308 °K; (Chamber temperature at 30 min)

T_2 = 273 °K; (Temperature at STP)

Step-3: Area occupied by chamber (A) in m^2 = 53 × 33/10000 (length x breadth / one hector (10000 m^2) = 0.1749 m^2

Step-4: CH_4 flux = $[\Delta X_{CH4} \times V_2 \times MW_{CH4}] / [T\times A]$

$= [1.289 \times 106128 \times 16 \times 1000 \times 60 \times 1]/ [1000000 \times 22400 \times 30 \times 0.1749]$

= 1.12 mg m^{-2} h^{-1}

Where, Time of gas collection/ flux capturing (T) = 30 minutes (flux time); Molecular weight of CH_4 (MW_{CH4}) = 16; Volume of gas at STP = 22400ml;

2. If the 0 and 30 minutes nitrous oxide (N_2O) concentration reading in gas chromatograph is 393.430 and 388.181 ppb then what is the flux of N_2O. The volume of manual chamber used in collection of gas samples from field was 124179 cm^3 [71 × 53 × 33 (height × length × breadth)], the final and initial temperature was 30 and 32 °C, there was no standing water in the field and 72 cm^3 is the shoot biomass volume covered for six rice hills inside the chamber and assume that the pressure inside the chamber was maintained as STP (760 mm) during the collection of gas.

Ans. We know that Flux= Change in concentration of gas / unit area/ unit time.

Step-1: Change in nitrous oxide concentration over 30 minute= $\Delta X\ N_2O$ (30 m reading – 0 minute reading) = 393.43 ppb – 388.181 ppb = 5.249 ppb

Step-2: Volume of gas chamber covered= [(H–h) × L × B] = 124179 cm^3

Where, [H (height of chamber) =71 cm, h (height of standing water) = 0 cm, L (length of the chamber) = 53 cm and B (breadth of the chamber) = 33 cm]

Effective box/ chamber volume (EBV) without temperature correction (V_1)

= [(H – h) × L × B]-V

= (124179 – 72) cm^3

= 124107 cm^3

Where, Volume covered by six rice hills inside the chamber (V) = 72 cm^3

Corrected chamber volume (CCV) (temperature correction) (V_2)

= $(T_2 \times V_1)/T_1$

= [(273 × 124107)/305]

= 111086 cm^3

[Where, T_1= 32°C + 273°K =305 °K; (Chamber temperature at 30 min)

T_2 = 273 °K; (Temperature at STP)

Step-3: Area occupied by chamber (A) in m^2 = 53 × 33/10000 (length × breadth / one hector (10000 m^2) = 0.1749 m^2

Step-4: N_2O flux = $[\Delta X_{N2O} \times V_2 \times MW_{N2O}] / [T \times A]$

= [5.249 × 111086 × 44 × 1000 × 60 × 1]/ [1000000 × 22400 × 30 × 0.1749]

= **13.10 μg m^{-2} h^{-1}**

Where, Time of gas collection/ flux capturing (T) = 30 minutes (flux time); Molecular weight of N_2O (MW_{N_2O}) = 44; Volume of gas at STP = 22400ml

3. If the 0 and 30 minutes carbon dioxide (CO_2) concentration reading in gas chromatograph is 418.328 and 403.345 ppm then what is the flux of CO_2. The volume of manual chamber used in collection of gas samples from field was 124179 m^3 [71 × 53 × 33 (height × length × breadth)], the final and initial temperature was 33 and 35°C, there was 2 cm standing water in the field and 72 cm^3/volume is the shoot biomass covered for six rice hills inside the chamber and assuming the pressure was constant during the collection of gas.

Ans.: We know that Flux= Change in concentration of gas / unit area/ unit time.

Step-1: Change in carbon dioxide concentration over 30 minute = ΔX_{CO_2} (30 m reading – 0 minute reading) = 418.328 ppm – 403.345 ppm = 14.983 ppm

Step-2: Volume of gas chamber covered= [(H-h) × L × B] = 120681 cm^3

Where, [H (height of chamber) = 71 cm, h (height of standing water) = 2 cm, L (length of the chamber) = 53 cm and B (breadth of the chamber) = 33 cm]

Effective box/ chamber volume (EBV) without temperature correction (V_1)

= [(H-h) × L × B]-V

= (120681– 72) cm^3

= 120609 cm^3

Where, Volume covered by six rice hills inside the chamber (V) = 72 cm^3

Corrected chamber volume (CCV) (temperature correction) (V_2)

= $(T_2 \times V_1)/T_1$

= [(273 × 120609)/308]

= 106903 cm^3

[Where, T_1 = 35°C + 273°K = 308 °K; (Chamber temperature at 30 min)

T_2 = 273 °K; (Temperature at STP)

Step-3: Area occupied by chamber (A) in m^2 = 53 × 33/10000 (length x breadth / one hector (10000 m^2) = 0.1749 m^2

Step-4: CO_2 flux = $[\Delta X_{CO2} \times V_2 \times MW_{CO2}] / [T \times A]$

= [14.983 × 106903 × 44 × 1000 × 60 × 1]/[1000000 × 22400 × 30 × 0.1749]

= **35.978 mg m^{-2} h^{-1}**

Where, Time of gas collection/ flux capturing (T) = 30 minutes (flux time); Molecular weight of CO_2 (MW_{CO2}) = 44; Volume of gas at STP = 22400 ml

4. If the mean methane fluxes at 30, 37, 43, 50, 57, 64, 71, 79, 86, 93, 101, 108, 115, 122 and 129 days after sowing are 0.946, 1.060, 1.070, 1.262, 1.412, 1.609, 1.630, 1.789, 2.431, 2.993, 4.341, 3.354, 2.323, 1.203 and 0.961 ppm, respectively. Then what is the seasonal CH_4 emission from the respective field estimated through weighted-average method. The duration of the crop is 150 days and it was transplanted at 21 days old seedling.

Ans. We have methane concentrations at different days of sampling and days after sowing (DAS).

Step-1: First date of sampling on 30th days after sowing, 2nd on 37 DAS and so on.

Here it is given 15 point of data, at different DAS, then we have to calculate the flux-differences at each consecutive days of sampling.

That is,

1st = 30 – 21=9 days; (Date of transplanting = 21 DAS)

2nd = 37 – 30= 7 days;

3rd = 43 – 37 = 6 days;

4th = 50 – 43 = 7 days;

5th = 57 – 50 = 7 days;

6th = 64 – 57 = 7 days;

7th = 71 – 64 = 7 days;

8th = 79 – 71 = 8 days;

9th = 86 – 79 = 7days;

10th = 93 – 86 = 7 days;

11th = 101 – 93 = 8 days;

12th = 108 – 101 = 7 days;

13th =115 – 108 = 7 days;

14th = 122 – 115 = 7 days;

15th = 129 – 122 = 7 days.

Step-2: (average (L18:2.323) × 7) + (average(2.323:1.203) × 7) + (average(1.203:0.961) × 7)) = 204.84 mg m^{-2} h^{-1}

Step-3: Seasonal CH_4 (kg ha^{-1}) = Mean emission (mg m^{-2} h^{-1}) × 10000 × 24/1000000

= 204.84 × 0.24 = 49.16 kg ha^{-1}

5. Find the volume of the two manual chambers one is for rice and another is for maize field. Where length, breadth and height of the chamber used in rice field is 53 cm, 33 cm and 71 cm, respectively and length, breadth and height of the chamber used in maize field is 65 cm, 41 cm and 100 cm, respectively.

   For rice field manual chamber

   Volume of chamber = [Height of the chamber × Length of chamber × Breadth of chamber = [71cm × 53cm × 33cm] = 124179 cm^3

   For maize field manual chamber

   Volume of chamber = Height of the chamber × Length of chamber × Breadth of chamber = [100cm × 65cm × 41cm] = 266500 cm^3

References

Bhatia A, Sasmal S, Jain N, Pathak H, Kumar R, Singh A. Mitigating nitrous oxide emission from soil under conventional and no-tillage in wheat using nitrification inhibitors. Agriculture, Ecosystems & Environment. 2010 Mar 15;136(3–4): 247–53.

Bhattacharyya P, Neogi S, Roy KS, Dash PK, Nayak AK, Mohapatra T. Tropical low land rice ecosystem is a net carbon sink. Agriculture, ecosystems & environment. 2014 May 1;189:127–35.

Bhattacharyya P, Roy KS, Neogi S, Dash PK, Nayak AK, Mohanty S, Baig MJ, Sarkar RK, Rao KS. Impact of elevated CO_2 and temperature on soil C and N dynamics in relation to CH_4 and N_2O emissions from tropical flooded rice (*Oryza sativa* L.). Science of the Total Environment. 2013 Sep 1;461:601–11.

Bhattacharyya P, Nayak AK, Mohanty S, Tripathi R, Shahid M, Kumar A, Raja R, Panda BB, Roy KS, Neogi S, Dash PK. Greenhouse gas emission in relation to labile soil C, N pools and functional microbial diversity as influenced by 39 years long-term fertilizer management in tropical rice. Soil and Tillage Research. 2013 May 1;129:93–105.

Bhattacharyya P, Roy KS, Neogi S, Chakravorti SP, Behera KS, Das KM, Bardhan S, Rao KS. Effect of long-term application of organic amendment on C storage in relation to global warming potential and biological activities in tropical flooded soil planted to rice. Nutrient cycling in agroecosystems. 2012 Dec 1;94(2–3):273–85.

Falge E, Baldocchi D, Olson R, Anthoni P, Aubinet M, Bernhofer C, Burba G, Ceulemans R, Clement R, Dolman H, Granier A. Gap filling strategies for defensible annual sums of net ecosystem exchange. Agricultural and Forest Meteorology. 2001 Mar 1;107(1):43–69.

Fejer MM, Magel GA, Jundt DH, Byer RL. Quasi-phase-matched second harmonic generation: tuning and tolerances. IEEE Journal of Quantum Electronics. 1992 Nov;28(11):2631–54.

Hutchinson GL, Mosier AR. Improved Soil Cover Method for Field Measurement of Nitrous Oxide Fluxes 1. Soil Science Society of America Journal. 1981;45(2):311–6.

IPCC, 2014: *Climate Change 2014: Impacts, Adaptation, and Vulnerability. Part A: Global and Sectoral Aspects. Contribution of Working Group II to the Fifth Assessment Report of the Intergovernmental Panel on Climate Change* [Field, C.B., V.R. Barros, D.J. Dokken, K.J. Mach, M.D. Mastrandrea, T.E. Bilir, M. Chatterjee, K.L. Ebi, Y.O. Estrada, R.C. Genova, B. Girma, E.S. Kissel, A.N. Levy, S. MacCracken, P.R. Mastrandrea, and L.L. White (eds.)]. Cambridge University Press, Cambridge, United Kingdom and New York, NY, USA, 1132 pp.

Moraes RC, Blondet A, Birkenkamp-Demtröder K, Tirard J, Orntoft TF, Gertler A, Durand P, Naville D, Begeot M. Study of the alteration of gene expression in adipose tissue of diet-induced obese mice by microarray and reverse transcription-polymerase chain reaction analyses. Endocrinology. 2003 Nov 1;144(11):4773–82.

Mukherjee R, Sarkar U. Development of a micrometeorological model for the estimation of methane flux from paddy fields: Validation with standard direct measurements. Environmental Modelling & Software. 2008 Oct 1;23(10–11):1229–39.

Mukherjee R, Barua A, Sarkar U, De BK, Mandal AK. Role of Alternative Electron Acceptors (AEA) to control methane flux from waterlogged paddy fields: Case studies in the southern part of West Bengal, India. International Journal of Greenhouse Gas Control. 2009 Sep 1;3(5):664–72.

Ruimy A, Jarvis PG, Baldocchi DD, Saugier B. CO_2 fluxes over plant canopies and solar radiation: a review. In Advances in ecological research 1995 Jan 1 (Vol. 26, pp. 1-68). Academic Press.

Webb EK, Pearman GI, Leuning R. Correction of flux measurements for density effects due to heat and water vapour transfer. Quarterly Journal of the Royal Meteorological Society. 1980 Jan 1;106(447):85–100.

Wilczak JM, Oncley SP, Stage SA. Sonic anemometer tilt correction algorithms. Boundary-Layer Meteorology. 2001 Apr 1;99(1):127–50.

Zhang R, van Hoek AN, Biwersi J, Verkman AS. A point mutation at cysteine 189 blocks the water permeability of rat kidney water channel CHIP28k. Biochemistry. 1993 Mar;32(12):2938–41.

5

Impact of Climate Change

5.1 Impact of climate change on agriculture

Climate change has both positive and negatively impact on agriculture, food security and land degradation. Climate change could exacerbate the land degradation processes by increases in rainfall intensity, drought frequency and severity, dry spells, flooding, heat stress, sea-level rise and permafrost thaw. The $1/4^{th}$ of ice-free land area in the world is prone to human-induced anthropogenic degradation (Certini and Scalenghe, 2011; Hooke et al., 2012). Soil erosion rate is 20-100 times faster than that of soil formation in agricultural field. The erosion rate is least is zero-tillage (non-disturbed field) and maximum at conventional intensive tillage practices. Climate change further exacerbates the land degradation. The permafrost areas, dry-lands, low-lying coastal regions, river deltas are more vulnerable. The occurrence of drought in dry-land was increased by around 1% per year during 1961-2013 periods. On an average 380-620 million people globally experienced desertification who lived in dry-land areas (1980s and 2000s) in South and East Asia, North Africa, Middle East, Sahara region, and Arabian peninsula. Moreover, people who are already living in degraded/ desertified regions are more negatively affected by climate change.

Specifically, the land surface air temperature has increased about twice as much as the global average temperature since the pre-industrial era. The land-surface air temperature has increased relatively higher than the world's average surface (land and ocean) temperature since the pre-industrial era (1850-1900). The average land-surface air temperature has enhanced by 1.53°C (1.38-1.68°C), whereas, world's surface (including land and ocean) air temperature has enhanced by 0.87°C (0.75-0.99°C). Subsequently, global warming triggered the frequency, intensity and duration of heat waves and related damage. The direct and indirect impacts of climate change on global agriculture is summarised below for an overall idea about the importance of the issues.

1. The drought events have increased in South America, Mediterranean region, west Asia, north-eastern Asia and Africa (during 1961-2013).

2. The intensity of heavy precipitation/ rainfall events has also increased in global scale.
3. 'Vegetation-Greening (VG)' was noticed (through satellite imagery) in South America, Europe, Asia, central of North America, and south-east Australia. The probable causes are extended growing season, CO_2 fertilisation and increased nitrogen deposition. However, 'Vegetation-Browning (VB)' was noticed in Central Asia, northern Eurasia, few parts of North America, may be due to water stress. But fortunately, the VG has observed over a larger area than VB, globally in last three decades.
4. The desertification has increased in Australia, parts of East and Central Asia and Sub-Saharan Africa during last 3-4 decades.
5. Specifically, in several lower-latitude areas, yields of some crops like, maize and wheat have decreased. However, the yields of maize, wheat and sugar beets have enhanced at higher-latitude in last three decades.
6. The lower animal productivity and growth rates were noticed in pastoral systems in Africa due to climate change consequences (1961-2013).

5.2 Impact of climate change on environment and food security

Globally, the expansion of arid-climate zones and contraction of polar-climate zones is much evident due to climate change and global warming in last 3-5 decades. Therefore, abundances and shifts in seasonal activities observed in related flora and fauna of those regions. Sea level rise has been causing higher coastal erosion and adding higher land-use pressure on coastal regions of Asia, Africa, Australia and South America. The fire major consequences of climate change imports on environment and food maturity are listed below:

1. Climate change consequences may lead to food insecurity and exhibits a complex interaction with food availability, food access, utilization and stability.
2. Protein content of crops, vegetables and fruits is affected negatively by higher atmospheric CO_2 concentrations.
3. Climate change would affect food chain supply through disruption of transport, manufacture, and retail, which may lead to limited food access.
4. The current food system accounts around 40% of total GHG emissions; this includes crop, livestock, from land use and land use change, deforestation and peat-land degradation, however, efficient food-supply system through reduction of food loss and waste could lowers the GHG emissions and also improves food security.

5. And finally, practices that create synergies between mitigation and adaptation to climate change would lead to low-carbon and climate-resilient pathways for ecosystem health and food security.

5.3 Predictions of climate change consequences on Agriculture

In boreal regions the tree line is expected to migrate towards north-ward and the growing season would be lengthens. The winter warming is predicted to be increased due to decreased snow cover and albedo. However, increased evapotranspiration may warm the growing season.

The predicted increased rainfall zone would trigger vegetation growth and consequently reduce regional warming. On the other hand, drier soil conditions resulted from climate change would increase the severity of heat waves.

Projected thawing of permafrost due to global warming is expected to increase the loss of soil carbon. However, during the 21st century higher vegetation growth in those regions may compensate the loss of soil carbon marginally.

In a trade-off mechanism, the desertification enhances global warming through the release of GHGs/CO_2 that also linked with the reduction of vegetation cover. At the same time, this decrease in vegetation cover tends to increase the local albedo that leading to surface cooling.

The urbanisation and global warming would heat-ups the cities and their surroundings. As per model predictions, the night-time temperatures would be affected more than that of day-time temperatures.

Climate change would impose higher stresses on land, livelihoods, biodiversity and human on some specific regions on 21st century which were not anticipated previously. For instance, the frequency as well as intensity of droughts is projected to increase in the southern part of Africa and Mediterranean region.

Current levels of global warming at around 1.5°C, the risks from dry-land water scarcity, permafrost degradation, wildfire damage, and food supply instabilities are projected to be high. However, warming at around 2°C the risk from permafrost degradation and food supply instabilities are projected to be very high. Further, at around 3°C warming risk from vegetation loss, wildfire damage, and dry-land water scarcity are also projected to be very high (IPCC 2014, 2018)

5.4 Climate change impacts on Indian agriculture

Indian agriculture is vulnerable to climate change primarily due to predominance (around 80%) of marginal and small farmers (up to 1 ha; and 1-2 ha land holding, respectively) and poor coping capacity. Moreover, most of the farming sectors are unorganized and heterogeneous. Above all, around 55-60% of India's net sown area is *rainfed* which are frequently negatively affected by abiotic stresses, like drought, flood, heat-waves, hailstorm, cyclones, submergence, salinity, etc. Specific impacts are listed below.

Crops, Livestock and Fisheries

1. Increase in land surface-temperature could reduce crop duration, alter photosynthesis efficiency, increases crop respiration rates. The virulence, survival and distributions of insect-pest populations also would be changed. Very frequently new equilibrium between pest-crops is developed under global warming.
2. Increase in air-soil temperature would increase evapotranspiration, hasten nutrient mineralization in soils and also reduces the fertilizer use efficiencies.
3. The indirect effects of climate change on Indian agriculture regulated by frequency and or intensity of seasonal floods and droughts and availability of irrigation water, water/soil erosion, and availability of energy.
4. The CO_2 fertilization could increase the yield of rice and wheat (C_3 plants) in coming few decades. However, in spite of this fertilization effect the wheat yield may likely to be reduced due to increased in respiration, decrease in crop growth duration, and reduction in water supply in north-western parts of India.
5. Changing rainfall pattern in monsoon, delayed monsoon, terminal draught and frequent water-stress causes reduction in crop yield in the *rainfed* areas due to increased crop-water demand.
6. Erratic seasonal behavior and global warming already deteriorate the quality of fruits, vegetables, tea, coffee, aromatics, and medicinal plants in many parts of India.
7. Sea level rise and coastal salinity intrusions in the parts of West Bengal, Odisha, Andhra Pradesh, Tamil Nadu and Andaban Nicobar Island would affect the crop yield in those regions.
8. In contrast, the decreased frost and cold waves in coming decades would lead to decrease the yield loss and frost damage to mustard and vegetables in northern India.

9. Climate change-induced enhanced temperature could increase the lignification in tissues of plant which would reduce the digestibility of fodder and feed to livestock.
10. Increased water scarcity would also decrease production and quality of food and fodder which could negatively affect the nutrition of livestock.
11. The changes of rainfall pattern in cooler region of India may outbreak the vector-borne diseases of livestock by the expansion of vector populations.
12. The demand for energy requirement, water and shelter would further increase under climate change scenario (global warming) for livestock in order to meet projected milk demands.
13. Climate change-induced heat stress is going to be negatively affected the performance of dairy animals of N-W and Central India.
14. The fish breeding, migration, harvests would likely to be affected by sea level rise in western coast of India.
15. Extreme cyclonic events both east and west coast of India is going to be negatively affected capture, production as well as marketing of marine fishes. Further, coral bleaching is likely to be increased due to higher sea surface temperature.

Soil and Water

1. There is high probability of reduction of soil organic carbon (SOC) in most of the agro-climatic regions of the country. Lowering of SOC would have direct negative effect on soil quality and soil health.
2. Immobilization of carbon in soil may enhance due to litter deposition of higher C:N ratio biomass from plants which have been progressively exposed to higher atmospheric-CO_2 concentration. Higher immobilization of carbon and relatively lower rate of decomposition of SOC would further reduce the soil-nutrient supply to plant.
3. Higher N mineralization with lower availability to plant would be in card in future climate change scenario (high soil temperature, quick wet-dry cycle) that subsequently leads to higher GHGs emission to atmosphere through volatilization and denitrification of nitrogen.
4. The soil erosion in hilly N-E and N-W region in India is unpredictable. However, higher rainfall volume and frequency and wind intensity may affect the severity and extent of erosions in hills.
5. Indian-Sundarban, Bengal and Odisha coast along with coastal regions of Andaman and Nicobar Island is vulnerable to climate change induced-sea level rise in coming 21st century.

6. Over exploitation of ground water due to increased irrigation demands, higher evapotranspiration losses and increased atmospheric temperature scenario in recent future may also result in lowering groundwater table in many parts of N-W and central India.
7. There would be short term increase of water flow in Ganges, Bhramaputra due to melting of glaciers but in the long run the availability of water would decrease considerably.
8. Flood frequency is expected to increase that leads to higher soil erosion in most parts of the country. The water balance and quality of groundwater (particularly in coastal region) in different parts of India is predicted to be disturbed due to erratic rainfall and intrusion of saline-sea water.

5.5 Key messages

The important messages from this small chapter are very clear that the impact of climate change on agriculture is both positive and negative. The CO_2 fertilization has positive impact on C_3 plants up to a certain level. However, increase of atmospheric and soil temperature has mostly negative effects on both crops and livestock. The impact of climate change on environments and food security is interlinked. Both demand-side and supply-side mitigation and adaptation strategies not only mitigate climate change but also ensure food security in future. Unfortunately, Indian agriculture is more vulnerable to climate change consequences due to small land holding, prevalence of economically weak farm families, lack of forecasting and insurance options available till date and top-to-bottom approach in policy making.

5.6 Probable questions

1. Explain the direct effect of climate change on world's agriculture.
2. What are the indirect impacts of climate change on agriculture?
3. Write the probable effects of climate change on future food security.
4. Why the Indian agriculture is more vulnerable to climate change?
5. Write the specific effects of climate change on livestock and aquaculture.
6. What are the probable positive effects of climate change on agriculture?
7. What might be the specific negative effects of climate change on soil and water systems?
8. How CO_2 fertilization and increasing atmospheric temperature are going to counter act the yield of cereals in coming future?
9. Write the possible negative effects of sea level rice both on agriculture and environments.

10. How the Indian coasts are going to be affected by predicted climate change?

References

Certini, G. and Scalenghe, R., 2011. Anthropogenic soils are the golden spikes for the Anthropocene. The Holocene, 21(8), pp.1269-1274.

Hooke, R.L., Martín-Duque, J.F. and Pedraza, J., 2012. Land transformation by humans: a review. GSA today, 22(12), pp.4–10.

IPCC: Intergovernmental Panel on Climate Change, 2018. Intergovernmental Panel on Climate Change. Global Warming of 1.5° C: An IPCC Special Report on the Impacts of Global Warming of 1.5° C Above Pre-industrial Levels and Related Global Greenhouse Gas Emission Pathways, in the Context of Strengthening the Global Response to the Threat of Climate Change, Sustainable Development, and Efforts to Eradicate Poverty. Rosenfeld, A., Dorman, M., Schwartz, J., Novack, V., Just, A.C. and Kloog, I., 2017. Estimating daily minimum, maximum, and mean near surface air temperature using hybrid satellite models across Israel. Environmental research, 159, pp.297–312.

6

Mitigation of GHGs Emissions and Climate Change

What is mitigation of GHGs emission? We must be clear about the answer of this fundamental question. In simple term, the mitigation of GHGs emission is the way or means by which the emission of those gases could be reduced. As we know from the previous chapters of this book that the GHGs emission could be occurred from natural as well as anthropogenic activities. Natural activities included volcanic eruptions, ocean current fluctuations, continental drifts which are responsible for emission of huge amount of CO_2 and other GHGs. The reduction of emissions from those natural activities is beyond our capabilities. However, the human induced emissions of GHGs emission could be reduced. We studied in previous chapters that fossil fuel burning, industrial and energy sector pollution, deforestation, land use change and agricultural could contribute to GHGs emission considerably. Mitigation techniques are different for each of those sectors that governed by lot of interrelated factors. In actual sense, the mitigation of GHGs emission should includes all the measures that can curtail the anthropogenic emission from all sectors. But that is out of the scope of this book. In this book, we specifically address the mitigation principles, options and efficiency of different techniques/ practices in agriculture only.

In that aspect, the mitigation of greenhouse gases emissions in agriculture is defined as "the means and ways by which emissions of GHGs could be reduced from agriculture". The ways and means actually referred to the techniques and practices and their amalgamation in agricultural sector both in production, processing and consumption aspects. So, mitigation is not only a techniques or combination of techniques it is actually an integrated approach for reducing greenhouse gases emission from agricultural systems. Primarily, there are two complementary approaches for mitigation of GHGs emissions from agriculture. Those two approaches are (i) supply-side and (ii) demand-side options or pathways. The supply-side approaches includes scientific land-use-changes, improvement of crop management, better livestock management practices, enhancing carbon storage in biota, soil carbon sequestration, significant substitution of fossil fuel by bio-fuels and use of renewable energy in agriculture. On the other hand, demand side approaches, taken in to account of shifting in the dietary pattern from non-vegetarian to vegetarian, curtail down the food loss and waste, and reducing the supply-chains (Figure 6.1). We can see, most of the approaches in supply-side are technology driven, while, approaches in demand-side are socio-economical and or policy driven. In this chapter we would thoroughly discuss the principles of mitigation, options of mitigation, mitigation cum adaptation approaches to reduce GHGs emission and climate change and few strategies to cope up with the vagaries of climate change.

6.1 Principles of mitigation of greenhouse gases emissions

The basic three principles of supply side-mitigation technologies for reducing GHGs emissions from agriculture are (i) reducing GHGs production in soil, (ii) retarding transformation and consequent emissions of those GHGs from soil to atmosphere, and (iii) increasing carbon sequestration in the soil-plant-microbes systems (Figure 6.2). The net GHGs exchanges and climate change feedback is actually resulted from the balance between production-transformation and consumption of GHGs in soil-plat-atmosphere continuum. Therefore, we should know the principles or mechanism for GHGs production-transformation-emissions in soil-plant-atmosphere system for effectively designing and executing the mitigation technologies. But it has to be remembered that the mitigation technologies are site specific and depends on various factors of agro-ecosystems.

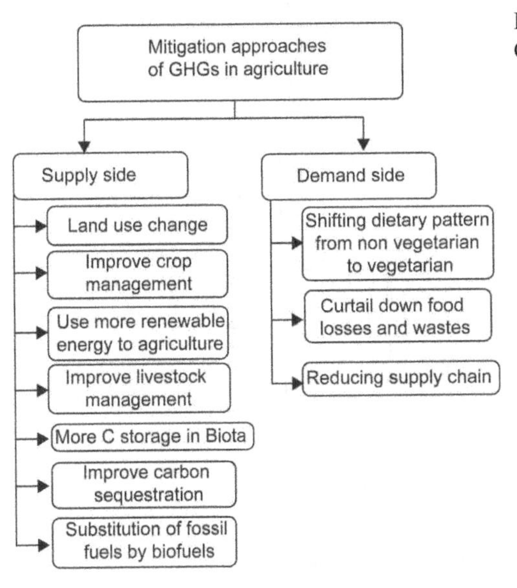

Figure 6.1 Mitigation approaches of GHGs emission in agriculture

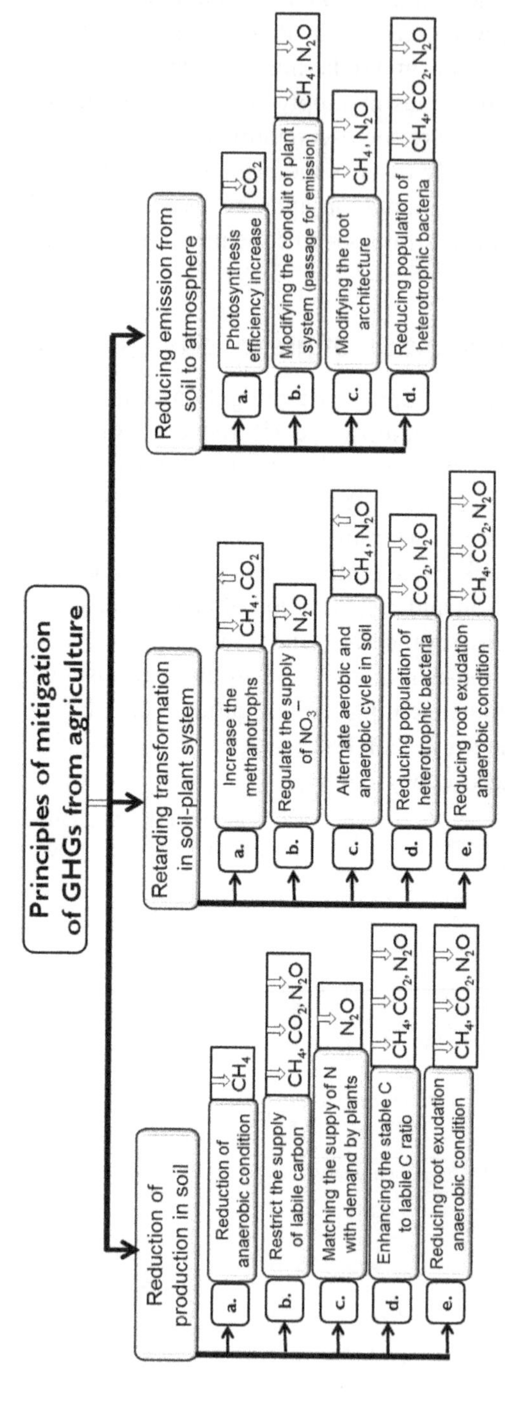

Figure 6.2 Principle of mitigation of GHGs emissions from agriculture

Methane, nitrous oxide and carbon dioxides are the major GHGs emitted from agricultural system. The principles of their production, transformation and emission in soil-plant system need to be understood.

6.1.1 Methane

The methane is mostly produced in rice paddy and wetlands. Its production and transformation are basically mediated by bio-chemical processes. Methane is primarily (85-90%) emitted from soil to atmosphere through the capillary tube-like aerenchyma tissues in rice plants which is plant mediated process. It also escapes from soil to atmosphere through diffusion, ebullition at water-air interface which are also governed by biochemical processes in soils. The details mechanism of that is already discussed in chapter 3 of this book. Wet lands or submerged rice soils provide congenial anaerobic condition for production of methane in soil with the help of methanogenic bacteria which utilizes easily decomposable labile carbon. Rice plants also provide a conduit to emit methane from soil to atmosphere. Therefore, it is clear that the three basic things are necessary for methane production in soil. Those are sufficient amount of labile (easily decomposable) carbonaceous substrate, anaerobic condition (~200mV) and the presence of methanogenic bacteria. Therefore, the mitigation principles of methane emissions in agricultural should address those three issues.

1. Regulation of labile organic carbon in the soil and reducing it supply to methanogens from soil systems.
2. Rhizosphere modification in the light of low rhizodeposition, low root-exudation and higher root-oxidase activities, so that methane production would be less and at the same time oxidation of CH_4 is taken place in rhizosphere.
3. Ensuring the alternate wetting and drying condition in the soil in order to keep soil redox potential above -200 mV.
4. Restrict the prolonged submergence condition in rice paddy (not more than 20 days at a stretch).
5. Maintaining relatively higher concentration of Mn^{+2}, Fe^{+3}, SO_4^{-2} ions, so that CH_4 production would be less in soil. This is because those ions act as favorable electron donors and not allow acetate or CO_2 to reduce and produce methane.
6. Keeping higher methanotrophic bacterial population over methanogens in order to favour methane oxidation over methane production.

6.1.2 Nitrous oxide

The trade-off mechanism exits between CH_4 and N_2O production and transformation in agricultural system. The trade-off refers that the soil condition which are favourable for methane production in not for nitrous oxide production and vice-versa. The strictly anaerobic condition required for methane production, while for production of nitrous oxide alternate oxidation-reduction cycle is required. The aerobic condition is necessary for nitrate (NO_3^-) production from ammonium (NH_4^+) and anaerobic condition is essential for conversion of N_2O /NO/N_2 from NO_3^-. The *Nitrosomonas* bacterial group are responsible for 1st step that is oxidation from NH_4^+ to nitrites (NO_2^-) and then *Nitrobacteria* species transformed NO_2^- to nitrates (NO_3^-). This oxidation process required aerobic condition. In the second step, in denitrification process, nitrous oxide is produced from nitrates by *Denitrifiers* and this process is anaerobic. The denitrification process is depending on the redox condition of the soil. Therefore, we can see that the conditions strictly favor for CH_4 production (detailed were discussed an earlier, chapter 3) not favorable for N_2O production and vice-versa.

The submerged rice-paddy system facilitates methane production while 'aerobic rice' and direct seeded rice favours nitrous oxide production. As in direct seeded upland ecology and 'aerobic rice' (grown rice as wheat), both aerobic and anaerobic condition prevails due to alternate wetting and drying which helps both oxidized and reduced nitrogen metabolism. However, the chemical, physical, and biological properties/ processes of rice-paddy and wetland affect the nitrogen transformation processes and N_2O emission in different ways than that of aerobic soil. So, nitrous oxide production requires the nitrogenous substrates in the form of ammonium ion for nitrification (where nitrites and nitrates and formed) and nitrate ions for denitrification. As the nitrates are soluble in water and any excess of that in soil-water system triggers the nitrous oxide production if the oxidation-reduction cycle continues in the system. Another important point to be remembered that soil system must have sufficient number of active nitrifiers and denitrifiers (microorganism). If any means, we could regulate the above three conditions, we can restrict the production, transformation and emission of nitrous oxide from agricultural production systems.

Now, we understood the basic mechanisms of N_2O production and transformation in agricultural soil. Based on those the principles of mitigation could be listed as,

1. Matching the demand for nitrogen (N) by plants to the supply of N through manure/ N-fertilizers.
2. Real-time and site-specific matching (demand-supply) is required, in order to increase the N use efficiency (NUE) and reduction of N losses including N_2O emission.

3. Inhibition of nitrification process and reducing solubilisation of N-substrates are other principles for mitigating N_2O emission.
4. Balanced supply of other nutrients like, phosphorus, potassium and micronutrients along with N, enhances NUE and thereby reducing N_2O emission.
5. Manipulation of forms and type of N-fertilizers for control/ slow supply N to soil.
6. Regulation of oxidation-reduction cycle in soil and maintaining proper balance among nitrosomonas, nitrobacters and denitrifiers in soil-water systems.
7. Supply and control of competitive electron donors and acceptors responsible for nitrous oxide production and subsequent transformation in soil system.

6.1.3 Carbon dioxide

Photosynthesis and respiration are the two fundamental processes regulate earth-atmosphere CO_2 balance. Terrestrial carbon budget in general and agricultural C-budget in particular is governed by the net CO_2 exchanges between earth-surface and atmosphere. Net ecosystem carbon dioxide exchanges are nothing but balance sheet of gross primary production (GPP) and ecosystem respiration (RE). The GPP represents the total assimilates as generated by photosynthesis, while, the RE refers to the respiratory CO_2 losses from organisms including the autotrophs and heterotrophs. Rhizosphere respiration represent the basal respiration from soil (primarily associated with soil organic carbon (SOC) decomposition) plus microbial and roots respiration.

On these basic backgrounds the principles of mitigation of CO_2 emission could be bulleted as,

1. Development of cultivars of high photosynthetic efficiency; that is breeding of varieties with higher capacity of harnessing atmospheric CO_2.
2. Increasing the below ground allocation of carbon through manipulation of plant-microbe interaction-mechanism.
3. Reducing the ration of labile carbon to stable carbon in soil.
4. Regulating heterotrophic activities in soil through controlled supply of labile carbon in the system and by soil amendments.
5. Reduction of soil priming effects and facilitating the carbon sequestration in soil-plant system.

6.2 Mitigation strategies of GHGs emissions

The mitigation strategies of GHGs emissions in agriculture could be categorized into two distinct pathways namely, supply-side and demand-side mitigation options. The supply-side mitigation strategies include, crop and livestock management, productive land use changes and carbon sequestration. Specifically, we can listed those as, manipulation of cropping sequence and systems, balanced manure-fertilizer application, introduction of deep rooted crops in up and medium land, biochar application, etc. Mitigation strategies in livestock sector must consider improved feeding and grazing for aquaculture and animal. Manure management and breeding of productive ruminant are also important. Social forestry and Agroforestry also should be considered as mitigation strategies in supply-side pathway; however, it is very often listed as adaptation strategies. In the demand-side mitigation strategies, reduction in food waste and loss, alteration in dietary pattern and shortening the supply-chain (Smith et al., 2014) are emphasized till now by FAO and IPCC (2018).

Agricultural production systems act both as source and sink of GHGs (Bhattacharyya et al., 2014, 2016). For instance, lowland rice ecology, mangrove ecosystem and wetlands act as CH_4 source but as CO_2 sink. However, aerobic-rice, pulse, and dry land-horticultural systems act as a CO_2 source. Crop management practices directly or indirectly mitigate GHGs emission from agriculture. Few of them directly curtail the emission, others increase the nutrient use efficiency and or productivity and by the way indirectly reduce the emission.

6.2.1 Supply side mitigation strategies

Smith et al. (2014) projected in IPPC 5th assessment report that the sum-total of mitigation potential from supply-side strategies is around 2.0-2.5 Gt CO_2-eq yr^{-1} by 2030. The other benefits of this approach; it can be directly applied to the field in the production system itself. The stakeholders like producers, farmers and retailers directly involved in those practices, so it is relatively easier to adapt. The details of each options/ strategies are discussed in coming sections.

Improved cultivars

Improved cultivars have the potential to curtail down the emission of GHGs could be introduced. As for example, development and introduction of rice cultivars having the potential to restrict the transport of CH_4 and N_2O from soil to atmosphere through modification of aerenchyma tissue-orientation. We know about 85-90% of CH_4/ N_2O emitted by oriented aerenchyma tissues in rice. The short duration upland/ aerobic rice cultivars (Kalinga 1) have higher CH_4 mitigation potential than long duration cultivars (Varshadhan) grown in deep water condition (Satpathy et al., 1998, Bhattacharyya et. al, 2019). The preferred ecology based selection of rice cultivars also helpful to mitigate CH_4 emission. In aerobic soil, the terminal heat-tolerant wheat variety P159-2/P157-8 and very short duration pulse cultivar (*cv.* Virat, for green gram) have the potential to curtail down the CO_2 and N_2O emission, due to very short duration and less nitrogen and water requirement, respectively.

Balanced application of manures and fertilizers

The CH_4 and N_2O emissions could be reduced by 15-20% with the deep placement of urea briquettes (big granules) (Figure 6.3) as compared to urea-broadcasting in rice (Chatterjee et al, 2016). Addition of potassium (K) and sulphate (SO_4^{-2}) in balanced proportion can reduce the CH_4 emission considerably from rice. This is because sufficient K in rice-rhizosphere increases the oxidizing power of roots and also increasing the iron oxidation which subsequently oxidize methane and reduces it emission. The sulphate competes with CO_2 and acetate for electron donor in submerged anaerobic soils, hence, reduces methane production. Therefore, addition of potassium sulphate (K_2SO_4) in rice-production systems (Wassmann et al., 1993) is a good strategy for reduction of GHG emission. In this connection, use of phosphorus or potassium or any micronutrients as sulphate fertilizer is a meaningful option for reduction of methane emission. Further, along with sulphate, iron and manganese also compete with acetate and CO_2 as an electron donor for methanogenesis (methane production) in soils. So any amendments /fertilizers/ manures generating those ions after dissolution in soil could reduce CH_4 production.

Apart from that balanced application of fertilizers also facilitates higher nutrient use efficiency by the crop and thereby reduces the losses including CH_4 and N_2O (gaseous losses).

Figure 6.3 Urea briquettes

Manures with a low C: N ratio (C: N <20:1) always preferred for reducing CH_4 emission as compare to manure/compost with high C: N ratio, as freshly prepared manures having higher C: N ratio favors methanogenesis in submerged soil. Substitution of fertilizer-N through manure (on equivalent-N basis) could reduce N_2O emission from the rice-systems as compared to application of inorganic fertilizer alone (Bhattacharyya et al., 2012).

Real time nitrogen management

Real time nitrogen management ensures supply of N from soil to plant at right time when crop demanded. This strategy could be implemented by using three basic techniques. First one, to apply the N-fertilizer at the critical physiological growth stages of crop. The second technique is use of leaf colour chart (LCC) for N and or SPAD meter which are based on greenness of the leaf. And the third one is sensor-based technology for identifying the nutrient deficiency of plants. All those techniques basically developed for increasing the NUE of crop. As those techniques reduce the losses of N and enhance the N-use efficiency, in turn automatically reduce the emission of CO_2 and N_2O. Encouragingly, crop specific LCCs have been developed for N-management in rice, wheat and maize. It is a cheap and easy adaptable technique that has the potential to reduce N_2O emission by 5-10 % (Bhatia et al., 2012; Mohanty et al., 2013).

Efficient and economic water management

Efficient water management is the key to agricultural productivity and so for reduction of GHGs emissions. It regulates the production, transformation as well as emission of GHGs in agricultural. Economics of water management is equally important as efficient management in recent time because of high value of water and its burgeoning scarcity. The direct seed rice (DSR) which promotes alternate wetting-drying cycle reduces CH_4 emission (20-30% as compared to transplanted rice-paddy), however, DSR facilitates N_2O emission if N is not managed properly. The simple and most effective practice in transplanted rice for reducing methane emission is mid-season drainage in between maximum tillering and panicle initiation stage of crop (reduction as high as 35-50%). Fertigation (Fertilizer mixing with irrigation water) and matching of N-fertilization schedule with irrigation is an effective strategy for reduction of N_2O emission in wheat, rice, maize, and sugarcane. Micro-irrigation techniques like, drip and sprinkler are recommended to high-value vegetable crops and in orchards for mitigating the N_2O and CO_2 emissions. Methods of irrigation like, increasing number of splits to sandy soils and check basin to orchards not only increase the water and N-use efficiency but at the same time reduces emission of N_2O.

Change in cropping system and cropping sequence

Ecology based efficient cropping sequence and cropping system significantly impacted the GHGs emissions. Introduction of pulse in rice based system particularly rice-rice system has the capacity to reduce the GHGs emission by 30-35% over rice-rice system in tropical lowland (Neogi et al., 2014). Rice-wheat system favoured N_2O emission (due to higher N-fertilizer use and alternate wetting-drying condition of the field), while rice-rice cropping sequence is responsible for higher CH_4 emission (due to continuous submergence). Maize-wheat favours higher CO_2 and N_2O emission, while any pulse based cropping system helps in reduction of nitrous oxide emission. Intensive cropping system (more than 2 crops annually) although provided higher economic return but at the same time causes higher GHGs emission. Therefore, tillage, cultural practice and real time fertilizer management are the nodal issues which have to be looked in to, for reducing GHGs emission from intensive cropping system.

Tillage and cultural practices

Zero and conservation tillage for reduction of CO_2 and N_2O emissions and carbon sequestration are the well established techniques. The zero and minimum tillage in DSR has already shown the mitigation potential of CH_4 and N_2O in rice-wheat cropping systems compared to rice-rice due to lower rate soil-C decomposition, oxidised top layers and relatively less rhizospheric effect. It was proved that DSR helps in curtail down of both the CH_4 and N_2O emission as compared to transplanted rice (Pathak and Aggarwal, 2012; Bhattacharyya et al., 2017). Recent time, the conservation agriculture encourages the use of the second-generation machinery like Zero-till seed drill, Happy seeder, etc., for *in-situ* residue management that have a good potential to reduce GHGs emissions in rice-wheat systems.

Nitrification inhibitors and slow release nitrogenous fertilizer

Nitrification inhibitors like, Nimin, Dicynamide (DCD), and Karanja oil restrict the first step of nitrification i.e. conversion of NH_4^+ to NO_2^- and there by reduces the emission of N_2O. The Nmin, a concentrated extract of seed of neem (*Azadirachta indica*) having 5% of neem-bitter acts as a nitrification inhibitor could be effectively used in rice-wheat system. The inhibitors specifically reduce the activities of *Nitrosomonas sps.* Slow release nitrogenous fertilizer like neem / sulphur coated urea significantly reduced the N_2O emission (5-10%) in rice and at the same time increased the NUE by 10% (Mazumdar et al., 2000). Those actually slow down the urea solubilization process in soil and check the N-losses. By this way those reduces the production and emission of N_2O from soil to atmosphere. Deep placement of urea super granules and urea briquettes in reduced soil-zone in another approach in lowland rice to curtail down N_2O emission by 20-25% (Chatterjee et al., 2016). Fly ash coated urea briquettes would be another option need to be validated in field condition.

Soil amendment and addition of biochar

Soil amendments refer to any substances that added to soil for improving soil physical and or chemical properties/ processes. There nature and characteristics varies with source material and condition at which those are used. Very often, soil amendments are used to reclaim the problem soils. Industrial by-product, geological substances and plant-derived materials are commonly used as soil amendments. For instance, industrial by-product like, phosphogypsum is used as a soil amendment to provide calcium and sulfur nutrition to crop and improving the soil physical condition of clay soils. The sulfate content in phosphogypsum in other way hinders methane formation in soil as by triggering the competition for substrates (hydrogen or acetate) among sulphur reducing bacteria and methanogens (Hussain et al., 2015). Basic slag, byproduct of steel industry, used as a soil amendment having silicate, free iron and manganese oxides, that act as electron acceptors, and an oxidizing agent (methane oxidation) in rice farming (Ali et al. 2008 and 2013) and thereby reducing methane emission. Biochar produced by pyrolysis process (burning in limited oxygen condition) add recalcitrant carbon in soil and helps in maintaining higher stable: labile carbon in soil which in turn reduces soil respiration and CO_2 emission (Kuzyakov et al. 2009).

Land use change- management

If you want to reduce the CH_4 emission instantly, you can convert wetland-agriculture to aerobic-agriculture. However, it is not a wise land use change. Wetland and lowland rice-paddy is a huge carbon sink (sink of CO_2 in particular). So, there is always a fear to loss a considerable amount of stored carbon wet/peat soil if you abruptly convert those to aerobic system. As an environmentalist point of view, I should recommend to restore the wet land as much as possible in order to sustain ecological balance. We should do intelligent land use management and discourage the conversion of mangrove, wetland, peat and bog lands to aerobic-agriculture or aquaculture.

Uplands should be used by cover crops, horticulture and pulses. Lowland for rice-paddy with proper drainage and real-time N management is ideal. In arid to semiarid zones having sandy soils should be discouraged for rice cultivation. Shifting cultivation with less frequency in hilly region must be discouraged.

Improved livestock management

Supply-side mitigation option for GHGs emission is incomplete without improve livestock management which contribute to 50% mitigation potential.

It primarily includes reduction in enteric methane-formation, curtail down N_2O emission through manure-management and pasture -manipulation. Those strategies have the potential to reduce GHG emission by 1.8 -2.4 Gt CO_2-eq yr^{-1} (Herrero et al., 2016). Feeding good quality diets like, legumes, industrially produced microbial protein, improved pasture sps., would help in reduction of GHGs emission per unit animal product (Pikaar et al., 2018). The other effective options for reducing GHG emission from livestock sectors are, promoting lower CH_4 producing breeds, improved feeding-system, introduction of dairy-crop mixed system for dual purpose (milk, crop, and meat), microbial-manipulation of rumen microflora, and use of feed additives (Hriston et al. 2013; ADAS, 2015). Recent approaches like, promotion of Hristov environment-friendly breeds, shifting from ruminants to pig or poultry (monogastrics) (Frank et al. 2018) are highly eco-friendly strategies for reducing the GHGs emissions from livestock sector. Manure management is equally important for mitigating GHGs emission, starting from feeding balanced diet to livestock, collection of excreta, storing in proper condition (to reduce CH_4 and N_2O emission) and delivery to the field (Bouwman et al. 2013, Bourdin et al., 2014). The most important source of methane i.e enteric fermentation in ruminant stomach through methanogenesis could be inhibited (30-75%; Hristov et al., 2015) by providing dietary lipid or anti-methanogenic agents like algae (*Asparogopsis taxiformis*), 3-nitro- oxypropanol (3-NOP).

Management of aquaculture and cellular agriculture

Aquaculture managements for GHGs mitigation mostly targeted on improvement of fish health, replacement of fish-based to crop-based ingredients and increment the feed conversion rates. The other options could be use of renewable energy and increase in input use efficiency (Barange at al. 2018). Cellular agriculture is the future of the livestock science. It includes in-vitro meat, cultured meat, hydroponic meat, synthetic meat, industrial production of milk, egg white and leather (Stephens et al., 2018). Those have huge GHGs emission mitigation potential in respect to reducing land use, ruminant CH_4 production and manures generated GHGs omission (Kumar et al. 2017).

Agroforestry

Agroforestry has good potential to reduce the emission of GHGs. It is equally important for crop and livestock sectors and as well as for both developed and developing countries. It primarily helps in carbon sequestration. Additionally improves soil organic matter build up (Mblow et al., 2014) and sustain better soil health. It helps in checking soil degradation through erosion and thereby reducing off-site carbon emission.

6.2.2 Demand-side mitigation strategies

Socio-economic factors include population growth rate, food-demand, policy-intervention and political stability drives the demand-side mitigation strategies. So obviously, those strategies are difficult to promote. However, IPCC (2014) predicts based on the preference of diet and region, a potential to mitigate GHGs emission through demand-side pathways is around 2.7-5.7 Gt CO_2-eq yr^{-1} by the year 2050. Three major drivers in this pathway are dietary preference, food loss and waste and length of the supply chain from producer to consumer.

Dietary pattern

The non-vegetarian diet causes higher GHGs emission than vegetarian diet. Among the non-vegetarian diet 'ruminant meat' generate highest GHGs emission. So, in simple language the plant-based diet has a lower GHG footprint as compared to animal-based diet (Nelson et al. 2016). Specifically, diet with beef and lamb has high negative impact on the environment. In this regard a flexitarian diet (limited meat + dairy) shown highest GHG mitigation potential followed by 'healthy diet' (limited sugar + meat+ dairy) and vegetarian over non-vegetarian (ruminant meat consumer) (Herrero et al., 2016). This mitigation potential was calculated on the basis of ruminant CH_4 reduction, land use change, area conversion, and production process-CO_2 emission. However, 'per day per capita calorie intake' is another approach to estimate mitigation potential of different diet-scenario (Smith et al., 2013).

Reduce the loss and waste of food

The loss and waste of food is responsible for 10% of total anthropogenic GHG emission (~ 44 Gt CO_2-eq yr^{-1}) in global level (Porter et al., 2016). Actually, a large amount of food is lost in developing countries (due to lack of storage facilities, mishandling and poor infrastructure), while, a large quantity of food is wasted in developed countries (due to lack of awareness and negligence) (Godfray et al., 2010). Therefore, infrastructure development in developing countries and consumer awareness in developed countries are the two key approaches to reduce losses and waste of food. As for example, about 20% of food produce was wasted in North America and Europe and 30% of food -produced was wasted in Africa during 2007 (FAO 2011). Specifically, 3% of global N_2O emission is contributed through the losses and waste of food in 2009 (Reay et al., 2012). However, the point to be remembered the largest losses of food occur as crop-biomass prior to harvest (Alexander et al. 2017). In low-income countries, majority of the food is lost during production, post-harvest handling, processing and transportation due to poor infrastructure facilities. These issues should be addressed by financial aid and positive policy framework. In developed countries smart packaging and processing technologies based on consumer preference should be encouraged (Wilson et al., 2017, FAO 2013) for mitigating GHGs emission.

Shortening the supply chain

Site specific production and marketing, based on local preference is the key to this mitigation strategy. Local level production and selling effectively check the losses of produce due to transportation and also reduces GHG emission associated with food-loss. However, we must see, in order to have an environmental friendly local food production-consumption system, we should not make an inefficient production system is long-run. For instance use of local unproductive breed may curtail the GHG emission by reducing the supply chain but at the same time imported animal products from the highly productive system may have much lesser GHG footprint. Therefore, the supply chain should be regulated on the basis of system productivity, transport facility and consumer preferences.

6.3 Adaptation strategies to climate change and mitigate GHGs emission

The term "adaptation strategies" to climate change broadly defined as the modification of non-biological as well as biological practices/ mechanisms that helps the system in general and agriculture (or say specific crop) to cope-up with the climate change-stresses. The cope-up mechanism includes crop survival, growth, development and propagation. However, there is a limit of adaptation to a specific system (or crop or organism) to what extent it could be modified or adjusted. We can categories the adaptation strategies in ways one for mitigating GHGs emissions and another for adaptation to climate change. As enhanced GHGs emission is the primary causes of climate change which manifested by higher temperature, sea level rise, higher cyclonic event, erratic precipitations, more salinity, etc. So, in broad sense the adaptation strategies must be same for climate change issues and mitigating GHGs emission, although we discussed those two in different subheading for better understanding.

6.3.1 Adaptation strategies to climate change

There are some inherent or we can say intrinsic adaptive capacity of crops and animals to cope up with higher temperature. However, some of those traits could also be induced to the organism through artificial breeding but the mechanism should be considered as inherent. The adaptive mechanism includes shifting of optimum thermal range, avoidance, escaping, thermal cooling, cutinization, stomatal closure, development of heat shock-protein, osmo-regulation, waxination, *etc.* are mostly considered as natural adaptation. In recent times breeding of crop varieties for heat tolerance through conventional as well as molecular techniques have been getting impetus. Those approaches generally started with screening of heat tolerant genotypes, followed by exploitation of desirable genes/QTLs and then introgression of genes those into new tolerant cultivars. Agro-physiological interventions are other effective tools to enhance adaptation ability like, change of dates of sowing, frequent irrigation, crop diversification etc.

Some of the strategies are listed below:

1. The easily adaptive crop management strategies to cope up climate change includes real-time adjustment of sowing time, modification of sowing method, contingent nursery, staggered nursery for rice, adjustment of nutrient and pesticide according to need.
2. Introduction of abiotic and biotic stress tolerant varieties in vulnerable areas.
3. Crop-diversification (with short and long duration crop; agronomic + horticultural crops) and integrated farming systems (both crop and animal component) are the most effective adaptive strategies to climate change vagaries. As those approaches have buffering capacity and flexibility that help the farmers not only to augment the production system but also to have improved and sustained income.
4. Perennial plantation with tolerant-horticultural crops (fruits, spices, timber crops etc.) along with in-built post-harvest management has higher adaptive capacity. Above all, the plantation crops and other agro-forestry systems also sequester carbon.
5. The three tier system is very effective for climate change adaptation. For instance, combination of banana (C_4 species), pineapple (CAM species) and *anthurium* (shade loving species) as intercrops in coconut (C_3 species) has found productivity and stable under climate change vagaries.

6.3.2 The adaptation strategies to mitigate GHGs emissions

We must know that when we are talking about adaptation strategies to mitigate GHGs emission from agriculture, it is very difficult to distinguish mitigation and adaptation strategies separately as both are complementary to each other. Therefore, in this section we discussed the adaptation cum mitigation strategies to reduce GHGs emission which are based on the mitigation principles described on section 6.1 and 6.2. The specific adaptation cum mitigation strategies for reducing CH_4, N_2O and CO_2 emissions are bulleted here for clear understanding.

1. Adapting rice-pulse cropping system instead of rice-rise to could reduce the CO_2 emission by 30-35% (Neogi et al., 2014).
2. Two to three days drainage in rice-paddy within the critical growth stages of maximum tillering and panicle initiation could reduce the CH_4 emission by 30-40% (Lu et al., 2000).
3. Retention followed by incorporation of rice straw at *kharif* season in lowland rice reduced the N_2O emission by 10%, but increased the CH_4 fluxes by 8% (Bhattacharyya et al., 2012). However, it should be encouraged as this technique also has carbon sequestration potential in medium to long run.
4. Phase wise conversion of conventional puddled rice zone to DSR belt with proper weed (through selective herbicides) and water management coupled with introduction of second generation farm machineries to curtail CH_4 as well as N_2O emissions (Pathak and Aggarwal, 2012).
5. In eastern and north-eastern India, transplanting of older seedling (21-30 days) of rice instead of young seedling (8-10 days) has the potential to reduce GHGs emissions.
6. Paired row cropping with rice-rice-*Sesbania* (row-orientation) and *in-situ* green manuring (i.e incorporation of *Sesbania* at 25-30 day after sowing) could be adapted for checking GHGs emission instead of traditional green manuring (incorporation of 40-45 days' old *Sesbania* before the transplanting of rice) (Bhattacharyya et al., 2014).
7. Early planting reduces emissions. Methane emissions were measured more in late-planted rice during the *kharif* season.
8. Frequent water-requirement based irrigation if adapted could reduce all the three important GHGs emission from agriculture.
9. Applications of phosphogypsum (2 t ha^{-1}), biochar (5 t ha^{-1}) and basic slag (1 t ha^{-1}), once in three years have the potential to curtail GHGs emission in rice production systems.
10. Selective application of butachlor, carbofuran, and hexachlorocyclohexane, have the potential to reduce CH_4 emission from agriculture by dropping the redox potential of the soil and inhibiting the methanogenic activities.
11. Adaptation of zero / minimum tillage reduces the N_2O as well CO_2 emissions compared to conventional tillage in the rice-wheat system.

6.4 Climate resilient agriculture (CRA)

'Climate Resilient Agriculture (CRA)' refers to the integration of mitigation, adaptation and real-time contingency practices in agriculture which increase the ability of the system to cope up various climate related stresses by resisting damage and recovering quickly. In brief, CRA means the ability of the agricultural system to bounce back to its original when subjected to certain climatic-stresses. Therefore, it is an intrinsic property of the system for recognizing the threats and also responses with higher degree of effectiveness. Here, it is important to know the stresses must be climate-change induced that include abiotic ones like, drought, flooding, heat/cold wave, erratic rainfall, long dry spells, salinity, extreme weather events, and biotic stress like, insect or pest population explosions, sudden weed infestation (new and invasive).

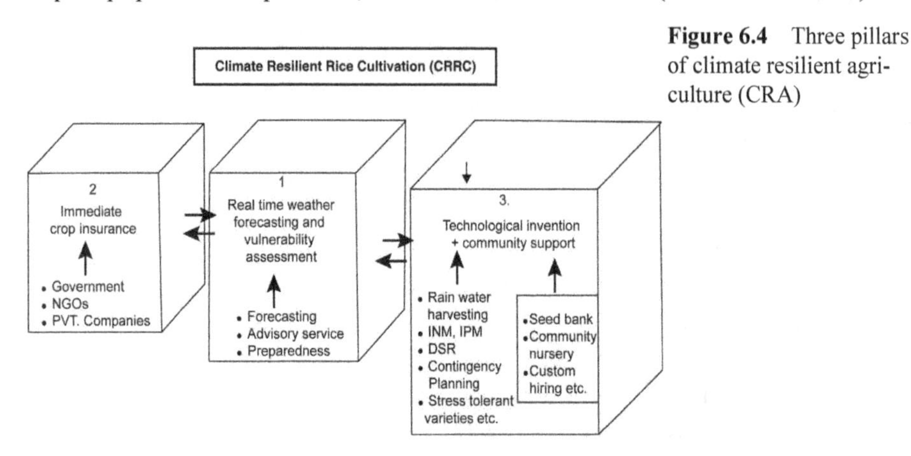

Figure 6.4 Three pillars of climate resilient agriculture (CRA)

Climate change has impacted the agricultural systems both directly and indirectly. The impacts could be experienced on quantity and quality of produce, pest-disease outbreak, resource utilization pattern, and on-time marketing. The primary aims of CRA are to increase resilience in the system through increasing adaptation capacities of all stakeholders (including farmer), sustaining natural resources and ensuring food security to climate change -challenges. The first pillar of CRA represents the real time weather forecasting and climate vulnerability assessment for agriculture (Figure 6.4). The real time crop-insurance and policy intervention is identified as the second pillar of CRA. Technological back-stepping and intervention is the third pillar of CRA. The third pillar includes the site specific technological interventions along with real time introduction of drought, heat and submergence tolerant crop varieties; adoption of location-specific conservation agriculture; and community support. The major technological interventions of CRA are as follows; (a) rain water harvesting and recycling; (b) conservation tillage; (c) DSR with mechanization; (d) crop residue management; (e) introduction of, flood, drought, submergence, heat-wave, salinity, and high temperature tolerant cultivars with package of practices; (f) contingency crop planning with community based support systems (custom hiring, staggered nursery for rice, seed bank). Further, social awareness towards climate variability, weather-based advisory services, preparedness for climatic extremes, and real-time crop insurance are the future of CRA which actually can make the system resilient.

6.5 Climate smart agriculture (CSA)

The 'Climate Smart Agriculture (CSA)' is actually CRA plus. Now question comes if CRA is well enough to address the climate change issues then why the approaches of CSA are evolved in recent times? The answer is not simple.

Now try to see the context critically, the world population is expected to enhance by 1.3 times by 2050 (9.7 billion in 2050), so to feed those many bellies, the agricultural production have to be increased by 60%. At the same time, climate change is a threat to food and nutritional security. Therefore, mitigation of GHGs emission and adaptation are also obligatory. And to answer those questions, the CSA approach evolved which would open up a window that simultaneously address the issues of increased food productivity, sustainability, resilience and food security of the system.

The history of CSA started in 2009. Then, the concept of CSA was got a shape in 1st global conference on 'Food Security, Agriculture, and Climate Change' in Hague, 2010. Subsequently, in 2012, at the 2nd global conference in Hanoi, Vietnam, the source book for CSA adopted. The next year, at 3rd global conference in Johannesburg, South Africa (2013), the alliance of CSA was discussed. Ultimately, at 'Climate Summit, in New York (2014)' the global alliance for CSA action plan was documented.

6.5.1 Definition and concept of CSA

Climate smart agriculture is defined as 'an integrated approach for developing technical, policy and investment conditions to achieve sustainable agricultural development for food security under climate change' (FAO, 2013). Like CRA it has also three pillars, namely, (i) sustainably increasing agricultural productivity and incomes, (ii) adapting and building resilience to climate change and (iii) reducing carbon emissions.

Actually, the concept of CSA is focusing on the food security in a changing climate. The 'CSA aims to improve food security, helps communities to adapt climate change and contributes to climate change mitigation by adopting appropriate practices, developing policies and mobilizing needed finances' (FAO, 2013). So, the dimensions of CSA are weather smart, water smart, , energy smart, nutrient smart, carbon smart and knowledge smart. The water smart practices in agriculture are DSR, raised bed for maize, drip irrigation for fruits and vegetable crops. Weather smart CSA includes robust weather forecasting with agro-advisory services, real time crop- livestock insurance and contingency seed supply system. Real time nitrogen management through LCC, SPAD meter based N-supply, sensor based nutrient application are components of nutrient-smart practices. Carbon and energy smart agriculture are primarily based on the 'conservation agriculture' and adaptation of zero tillage. Last but not the least knowledge smart refers to the capacity building of youths, using the ICTs and practicing the gender equality.

6.5.2 Approaches of CSA

The CSA is integration of three dimensions of sustainable development, namely; economic, social and environmental by addressing the issues of both 'food security and climate challenges'. The practices, policies and institutions must be worked hand on hand to get balance synchronization of the agricultural components like, soil, water, energy, crop and livestock (Figure 6.5). In soil and water management conservation agriculture (CA) is an established option. The CA integrates the components like minimum soil disturbance, residue retention and crop diversification. The other management options are, rain water harvesting, alternate wetting drying, contour bunding, micro-irrigation (drips and sprinkler) etc. The crop management in CSA primarily focused on profitability, stress tolerance cultivar development, legume-based crop rotations, and improved storage and processing facilities. The livestock management considers improved feeding strategies to cattle; introduction of rotational grazing in pastures; good quality fodder, conservation and restoration of grassland; and better shelter for livestock.

6.6 Key messages

1. The mitigation of GHGs emission is the means/techniques/technologies/ practices by which the emissions of those gases are reduced.
2. In agriculture; the mitigation of GHGs emissions refers to the means and ways by which emissions of GHGs could be reduced from agriculture.
3. There are two approaches of mitigation of GHGs in agriculture. They are supply-side and demand-side options or pathways.
4. The supply-side approaches includes scientific land-use-changes, improvement of crop management, better livestock management practices, enhancing carbon storage in biota, soil carbon sequestration, significant substitution of fossil fuel by bio-fuels and use of renewable energy in agriculture.

5. The demand-side approaches, includes the shifting in the dietary pattern from non-vegetarian to vegetarian, curtail down the food loss and waste, and reducing the supply-chains.
6. The basic three principles of supply side-mitigation approaches for reducing GHGs emissions from agriculture are (a) reducing GHGs production in soil, (b) retarding transformation and consequent emissions of those GHGs from soil to atmosphere, and (c) increasing carbon sequestration in the soil-plant-microbes systems
7. The supply-side GHGs emissions mitigation strategies include, crop and livestock management, productive land use changes and carbon sequestration.
8. The manipulation of cropping sequence and systems; balanced manure-fertilizer application; introduction of deep rooted crops in up and medium land; biochar application; improved feeding and grazing for aquaculture and animal; manure management and breeding of productive ruminant; social forestry and Agroforestry are the specific supply-side GHGs emissions mitigation techniques/ technologies/practices.
9. The adaptation strategies to climate change are defined as the modification of non-biological as well as biological practices/ mechanisms that helps the system in general and agriculture to cope-up with the climate change-stresses.
10. The easily adaptive crop management strategies to cope up climate change includes real-time adjustment of sowing time, modification of sowing method, contingent nursery, staggered nursery for rice, adjustment of nutrient and pesticide according to need.
11. The Climate Resilient Agriculture (CRA) is defined as the integration of mitigation, adaptation and real-time contingency practices in agriculture that increases the ability of the system to cope up various climates related stresses by resisting damage and recovering quickly.
12. In nut shell the CRA means the ability of the agricultural system to bounce back to its original when subjecting to certain climatic-stresses.
13. There are three pillars of CRA. They are (a) real time weather forecasting and climate vulnerability assessment for agriculture, (b) real time crop-insurance with policy intervention and (c) the technological back-stepping and intervention.
14. The major technological interventions of CRA are (a) rain water harvesting and recycling; (b) conservation tillage; (c) DSR with mechanization; (d) crop residue management; (e) introduction of, flood, drought, submergence, heat-wave, salinity, and high temperature tolerant

cultivars with package of practices; (f) contingency crop planning with community based support systems.

15. The Climate Smart Agriculture (CSA) is defined as an integrated approach for developing technical, policy and investment conditions to achieve sustainable agriculture; (i) sustainably increasing agricultural productivity and incomes; (ii) adapting and building resilience to climate change and (iii) reducing carbon emissions.
16. The dimensions of CSA are weather smart, water smart, , energy smart, nutrient smart, carbon smart and knowledge smart.

6.7 Probable questions

1. What are the supply-side and demand-side options of GHGs mitigation in earth- atmosphere?
2. Write the basic principles of mitigation of GHGs in agriculture?
3. What is the trade-off between mitigation of methane and nitrous oxide emission in agriculture?
4. Explain 5 mitigation strategies of each for CH_4 and N_2O emission from agriculture?
5. Real time nitrogen management is the key to reduction of N_2O emission and increasing NUE in agriculture: Explain.
6. Write adaptation strategies to climate change stresses?
7. What are the adaptation options for reducing GHGs emission from agriculture?
8. Explain four adaptations cum mitigation strategies for reducing GHGs emission in agriculture.
9. Define climate resilient agriculture and explain its basic concept.
10. Distinguish the CRA and CSA.
11. Write the need of developing CSA approaches.
12. Explain the dimensions of CSA.

References

Aggarwal, P., Pathakl, H., Kumarl, S. N., Sharma, P., & Vishwavidyalaya, H. K. 2012. South Asia perspectives on climate change and agriculture: Adaptation options. Handbook of Climate Change and Agroecosystems: Global and Regional Aspects and Implications. 2, 209.

Alexander, P., C. Brown, A. Arneth, J. Finnigan, D. Moran, and M. D. A. A. Rounsevell, 2017: Losses, inefficiencies and waste in the global food system. Agric. Syst.,

153, 190–200, doi:10.1016/j.agsy.2017.01.014. http://dx.doi.org/10.1016/j agsy.2017.01.014.

Ali, M.A., Hoque, M.A., Kim, P.J., 2013. Mitigating global warming potentials of methane and nitrous oxide gases from rice paddies under different irrigation regimes.AMBIO 42, 357–358.

Ali, M.A., Lee, C.H., Kim, P.J., 2008. Effect of silicate fertilizer on reducing methane emission during rice cultivation. Biol. Fertil. Soils 44, 597–604.

Barange, M., T. Bahri, M. C. M. Beveridge, K. L. Cochrane, S. Funge-Smith, and F. Poulain, 2018: Impacts of climate change on fisheries and aquaculture: synthesis of current knowledge, adaptation and mitigation options. FAO Fisheries and Aquaculture Technical Paper No. 627. Rome, Italy, 628 pp.

Bhatia, A., Pathak, H., Jain, N., Singh, P.K. and Tomer, R., 2012. Greenhouse gas mitigation in rice–wheat system with leaf color chart-based urea application. Environmental monitoring and assessment, 184(5), pp.3095-3107.

Bhattacharyya, P. and Pathak, H., 2017. Climate resilient agriculture: impacts and options for adaptation in India. SATSA Mukhaptra Annual Technical Issue, 21, pp.35-45.

Bhattacharyya, P., Neogi, S., Roy, K.S., Dash, P.K., Nayak, A.K. and Mohapatra, T., 2014. Tropical low land rice ecosystem is a net carbon sink. Agriculture, ecosystems & environment, 189, pp.127-135.

Bhattacharyya, P., Roy, K.S., Nayak, A.K., 2016. Greenhouse Gas Emmission from Agriculture: Monitiring, Quantification & Mitigation. Narendra Publishing House. New Delhi. ISBN 13-9789384337964.

Bhattacharyya, P., Roy, K.S., Nayak, A.K., Shahid, M., Lal, B., Gautam, P. and Mohapatra, T., 2017. Metagenomic assessment of methane production-oxidation and nitrogen metabolism of long term manured systems in lowland rice paddy. Science of the Total Environment, 586, pp.1245-1253.

Bhattacharyya, P., Roy, K.S., Neogi, S., Adhya, T.K., Rao, K.S. and Manna, M.C., 2012. Effects of rice straw and nitrogen fertilization on greenhouse gas emissions and carbon storage in tropical flooded soil planted with rice. Soil and Tillage research, 124, pp.119-130.

Bhattacharyya, P., Dash, P.K., Swain, C.K; Radhy, S.R., Roy, K.S., Neogi S., Berliner, J.; Adak, T., Pokhare, S.S., Baig, M.I. and Mohapatra, T., 2019. Mechanism of plant mediated methane emission in tropical low land rice. Science of the Total Environemnt, 651, pp. 84–92.

Bourdin, F., R. Sakrabani, M. G. Kibblewhite, and G. J. Lanigan, 2014: Effect of slurry dry matter content, application technique and timing on emissions of ammonia and greenhouse gas from cattle slurry applied to grassland soils in Ireland. Agric. Ecosyst. Environ., 188, 122–133, doi:10.1016/j.agee.2014.02.025.

Bouwman, L., and Coauthors, 2013: Exploring global changes in nitrogen and phosphorus cycles in agriculture induced by livestock production over. Proc Natl Acad Sci, 110, 20882–20887, doi:10.1073/pnas.1206191109.

Chatterjee, D., Mohanty, S., Guru, P.K., Swain, C.K., Tripathi, R., Shahid, M., Kumar, U., Kumar, A., Bhattacharyya, P., Gautam, P. and Lal, B., 2018. Comparative assessment of urea briquette applicators on greenhouse gas emission, nitrogen loss and soil enzymatic activities in tropical lowland rice. Agriculture, ecosystems & environment, 252, pp.178-190.

Godfray, H. C. J., and Coauthors, 2010: Food security: the challenge of feeding 9 billion people. greenhouse gas emissions from food loss & waste in the global food supply chain. Sci. Total Environ., 571, 721–729, doi:10.1016/j.scitotenv.2016.07.041.

Goose, R.J. and Johnson, B.E. 1993. Effect of urea pellet size and dyciandiamide on residual ammonium in field microplots. Comm.Soil Sci. Plant Anal. 24: 397-409.

Herrero et al., 2016. Greenhouse gas mitigation potentials in the livestock sector. Nat. Clim.15 Chang., 6, 452–461, doi:10.1038/nclimate2925. 16.http://www.nature.com/doifinder/ 10.1038/nclimate2925.

Hristov, A. N., and Coauthors, 2013a: SPECIAL TOPICS — Mitigation of methane and nitrous oxide emissions from animal operations: III . A review of animal management mitigation options. J. Anim. Sci., 91, 5095–5113, doi:10.2527/jas2013-6585.

Hristov, A. N., and Coauthors, 2015: An inhibitor persistently decreased enteric methane emission fromdairy cows with no negative effect on milk production. Proc. Natl. Acad. Sci., 112, 10663–10668.

Kumar, Y., R. Berwal, A. Pandey, A. Sharma, and V. Sharma, 2017: Hydroponics meat: An envisaging boon for sustainable meat production through biotechnological approach - A review. Int. J. Vet. Sci. Anim. Husb., 2, 34–39. http://www.veterinarypaper.com/archives/2017/2/1/A/1-3-25

Kuzyakov, Y., Subbotina, I., Chen, H., Bogomolova, I. & Xu, X., 2009. Black carbondecomposition and incorporation into soil microbial biomass estimated by ^{14}C labeling. Soil Biol. Biochem. 4 1, 210 – 219.

Lu, W.F., Chen, W., Duan, B.W., Guo, W.M., Lu, Y., Lantin, R.S., Wassmann, R., Neue, H.U., 2000. Methane emission and mitigation options in irrigated rice fields in southeast China. Nutrient Cycling in Agroecosystems 58, 65–74.

Majumdar, D., Kumar, S., Pathak, H., Jain, M.C. and Kumar, U., 2000. Reducing nitrous oxide emission from an irrigated rice field of North India with nitrification inhibitors. Agriculture, Ecosystems & Environment, 81(3), pp.163-169.

Mohanty, S., Nayak, A.K., Kumar, A., Tripathi, R., Shahid, M., Bhattacharyya, P., Raja, R. and Panda, B.B., 2013. Carbon and nitrogen mineralization kinetics in soil of rice–rice system under long term application of chemical fertilizers and farmyard manure. European Journal of Soil Biology, 58, pp.113-121.

Nelson, M. E., M. W. Hamm, F. B. Hu, S. A. Abrams, and T. S. Griffin, 2016: Alignment of Healthy Dietary Patterns and Environmental Sustainability: A Systematic Review. Adv. Nutr. An Int. Rev. J., 7, 1005–1025, doi:10.3945/an.116.012567

Neogi, S., Bhattacharyya, P., Roy, K.S., Panda, B.B., Nayak, A.K., Rao, K.S. and Manna, M.C., 2014. Soil respiration, labile carbon pools, and enzyme activities as affected by tillage practices in a tropical rice–maize–cowpea cropping system. Environmental monitoring and assessment, 186(7), pp.4223-4236.

Pathak, H. and Aggarwal, P.K., 2012. Low carbon Technologies for Agriculture: A study on Rice and wheat Systems in the Indo-Gangetic Plains. Indian Agricultural Research Institute, p. xvii, 78.

Pikaar, I., S. Matassa, K. Rabaey, B. Laycock, N. Boon, and W. Verstraete, 2018: The Urgent Need to Re-engineer Nitrogen-Efficient Food Production for the Planet. Managing Water, Soil and Waste Resources to Achieve Sustainable Development Goals, Springer International Publishing, Cham, 7 35–69 http://link.springer.com/10.1007/978-3-319-75163-4_3.

Porter, S. D., D. S. Reay, P. Higgins, and E. Bomberg, 2016: A half-century of production-phase Production of Second Generation Biofuels: Trends and I nfluences. 16pp. http://www.dovetailinc.org/report_pdfs/2017/dovetailbiofuels0117.pdf.

Reay, D. S., E. A. Davidson, K. A. Smith, P. Smith, J. M. Melillo, F. Dentener, and P. J. Crutzen, 2012: Global agriculture and nitrous oxide emissions. Nat. Clim. Chang., 2, 410–416,

Satpathy, S.N., Mishra, S., Adhya, T.K., Ramakrishnan, B., Rao, V.R. and Sethunathan, N., 1998. Cultivar variation in methane efflux from tropical rice. Plant and Soil, 202(2), pp.223-229. Science, 327, 812–818, doi:10.1126/science.1185383.

Smith et al., 2013: How much land-based greenhouse gas mitigation can be achieved without compromising food security and environmental goals? Glob. Chang. Biol., 19, 2285–2302, doi:10.1111/gcb.12160.

Smith, P., and Coauthors, 2014: Agriculture, Forestry and Other Land Use (AFOLU). Climate Change 2014: Mitigation of Climate Change. Contribution of Working Group III to the Fifth Assessment Report of the Intergovernmental Panel on Climate Change, O. Edenhofer et al., Eds., Cambridge University Press, Cambridge, United Kingdom and New York, NY, USA.http://www.ipccnggip.iges.or.jp/public/2006gl/pdf/4_Volume4/V4_04_Ch4_Forest_Land.pdf.

Wassmann, R., Papen, H. and Rennenberg, H., 1993. Methane emission from rice paddies and possible mitigation strategies. Chemosphere, 26(1-4), pp.201-217.

7

Putting Together and Way Forward

7.1 Economics and Valuation of GHGs emissions

Since the late 20th century, it has been largely accepted that cause of high concentration of GHGs in atmosphere is induced by human activities. Greenhouse gas emissions and related climate change has opened up debate on overall economic losses and ways to reduce these loses. Moreover, apart from curbing monetary losses; the global political leaders as well as reputed economists are looking towards making money by saving the emissions. Recently, the researchers have started to emphasize on the actual value added to the ecosystem by curbing emissions, discrete of the economic benefits. This section will discuss the economics and ecosystem services related to the management of GHG emissions.

Origin

The initiative to curb the emissions started in 1992 with the United Nations Framework Convention on Climate Change (UNFCCC) at the Earth Summit in Rio de Janeiro. The Convention took place from 3 to 14 June, 1992. The UNFCCC is an international environmental treaty designed to "stabilize greenhouse gas concentrations in the atmosphere at a level that would prevent dangerous anthropogenic interference with the climate system". The UNFCCC had non-binding limits on greenhouse gas emissions for individual countries and the framework was only meant for formulating precise international treaties (called "Protocols" or "Agreements"). It contained no execution mechanisms, however, it considered some negotiations and suggestions to formulate the action plans and regulation guidelines under UNFCCC to fulfill the objective of reduction of GHG emissions to the desired level.

In 1997, leaders of 180 countries signed a protocol in Kyoto, Japan, to take immediate action against climate change, famously known as Kyoto protocol. The Kyoto Protocol is extended form of the UNFCCC, which recognized the contribution of developed countries to greenhouse gas emissions. Therefore, the responsibilities for the GHGs reduction in the atmosphere were made legally binding. Article 17 of the protocol acknowledges that individual countries can reduce the emissions in different capacities, owing to economic development or advanced technologies. It was also recognized that industrialized countries were mainly responsible for human induced GHGs omissions. Therefore, the Kyoto protocol called for commitments from 37 industrialized countries to bring down the GHGs levels to the levels that are 5.2% less than those of 1990. The emission level in 1990 has been considered as baseline emission. Baseline emissions refer to the production of greenhouse gases that have occurred in the past and which are being produced prior to the introduction of any strategies to reduce emissions. This historical measurement acts as a benchmark to evaluate the extent of emission reduction after making a subsequent effort technologically or through change in policies or by any other means. The period of action was marked from the years 2008 to 2012. The 37 countries included, Australia, the European Union (and its 28 member states), Belarus, Iceland, Kazakhstan, Liechtenstein, Norway, Switzerland, and Ukraine. Under the Kyoto Protocol, six GHGs were listed in Annex A, viz., carbon dioxide (CO_2), methane (CH_4), nitrous oxide (N_2O), hydrofluorocarbons (HFCs), perfluorocarbons (PFCs), and sulphur hexafluoride (SF_6).

Mechanism for achieving this target

Three flexible mechanisms were formulated, also known as Flexibility Mechanisms or Kyoto Mechanisms. The three mechanisms involved (i) Emissions Trading, (ii) Clean Development Mechanism and (iii) Joint Implementation. These are mechanisms under the Kyoto Protocol intended to lower the overall costs of emitting less (emitting less incurs higher costs). These mechanisms enable parties/ countries to achieve emission reductions or to remove carbon from the atmosphere in other countries to make the process cost-effective. For example, cost of reduction of one unit of emission is much higher in a developed country like United States of America or UK because these countries are highly mechanized/ industrialized. In the developing counties like India, the production of many commodities engages human labour (human labour is relatively cheaper in India) with mush lesser emissions. So, a developed country can meet the target by producing a certain commodity using human labour in these developing countries and meeting their targets in terms of cuts in the cost of production as well as the emissions. The developing country is also expected to benefit directly from the employment generation in the country. Indirectly, it can also help on cutting down the imports of certain commodities as they are produced domestically, easing the economy. In the long run, the domestic markets may flourish with more and more commodities being produced within the country. The cost of reducing or regulating emissions varies considerably from country to country, but any reduction is good for the environment and it contributes to the overall healthy global atmosphere. The US government was the first to introduce flexibility mechanisms in Kyoto Protocol which was hailed by developed countries but highly criticized by the developing countries. The developing countries were uncertain about the implementation and the economic and cultural consequences. Poor regulation could easily overtake and may even crush the domestic industries in the developing countries. However, concerted efforts were made to negotiate over the Kyoto Protocol and finally, the developing countries agreed to the mechanisms.

7.1.1 Emissions trading

The fundamental concept of emissions trading was to remunerate the emissions saved or capped in any industrial activity. Later it was also named as cap and trade or carbon trading. The idea was to control GHG emission by providing economic incentives for achieving emissions reductions. The countries are required to pay for the extra emissions over the permissible limits (thus encouraging an emission cut) or selling the spare emission units (emissions permitted to them but not used). The emissions can be reduced or spared, for example, by replacing less efficient system (eg. equipment consuming huge quantity of fuel) by a highly efficient system (eg. equipment consuming least quantity of fuel). Thereby, a new commodity in the form of emissions and a new market in the form of carbon market were created. The CO_2 being the major GHG constituting the bulk; the emissions trading became synonymous with carbon trading. Emissions trading is calculated and expressed in tonnes of carbon dioxide equivalent or tCO_2e.

After the Kyoto Protocol in 1997, several amendments were made in due course of time. In 2010, UNFCCC stated that global warming should be limited to below 2.0°C (3.6 °F) compared to the levels of pre-industrial era. In the Doha Amendment, 2012, the period of action was revised from 2013 to 2020. The Paris Agreement was adopted in year 2015 (entered into force from November 2016), which considered emission reductions from 2020. The commitments by the countries were made under Nationally Determined Contributions, lowering the target to 1.5°C.

How emission/ carbon is traded?

Firstly, a restriction (quantitative limit) on the emissions (emissions cap) by emitters is fixed on the basis of baseline emissions. The government recognizes/ fixes an emissions cap and issues a quantity of emission allowances consistent with that cap. The country or the entity (industry, company, organization etc.) may buy and sell allowances, at the established market price for emissions. Any polluter entity or country (which emits higher levels of GHGs than permitted) has the right to emit only after paying for the emission that are above the permissible limits. Under carbon trading, the country or entity having fewer emissions i.e. emissions below the emission cap, can earn revenues by selling the spare carbon to the polluter country or entities. So, the revenue for any given expenditure on carbon reduction, can come from the polluter entities. That means, the polluter pays to the non-polluter. The emissions trading follows the economic basis similar to the concept of property rights.

Costs and valuation of emissions

Emissions trading is calculated and expressed in tonnes of carbon dioxide equivalent or tCO_2e. Many years after the Kyoto Protocol, the details of the mechanism of fixing costs and valuation are still being negotiated. In general, the polluting entity has to pay for inputs directly related to emission, for e.g. the costs of the fuel being used (the polluter pays). There are arguments in favour of considering all the social costs (human health impact due to global warming) involved, which will not only take all the costs into account but also influence the decisions and actions of the polluting entity. The logic is sound, however, the calculation of the social costs is not simple as it would need to consider the value of future climate impacts. Valuations can be difficult since not all goods and services have a market price.

Numerous methods are being used to deduce the prices for non-market goods and services; the pricing of many other goods and services are still in development. These valuations are based on many assumptions and keep changing with place and time which make the setting of prices difficult. The valuations of climate-change impacts on human or animal health are an example of that. The tolerance to different health conditions vary widely among people depending on the genetic and environmental make up. The extent of damage caused to the human or animal health due to change in climate will also vary widely. A single value for damage equivalent in monetary terms may not work for all the regions, and the results could be highly misleading. The potential positive benefits from climate change in particular regions, e.g., tourism, may or may not negative impacts in other regions, and vice-versa.

In spite of all these anomalies, economic analysis allows a consistent treatment of climate change impacts. It also gives an idea of the possible impact of climate change, and the policy makers can use some of these information and make policies accordingly to tackle the unseen impacts.

Present status of emission trade

About 50 countries have applied a price on carbon saving. Apart from the EU scheme which is the biggest initiative presently and China has implemented eight pilot projects. If the pilot projects succeed, it will be the largest national trading system with regard to carbon in the world. Like China, some other countries have shown interest in internal carbon pricing. Thousands of big and small businesses have agreed for a comprehensive common pricing on carbon. Many top executives went on to join the World Economic Forum's CEOs, climate leaders and the Carbon Pricing Leadership Coalition continues to encourage carbon trading. Over 1200 registered companies have made assessment on the climate risks of their businesses. This may help them in decision making in the long run. Task Force on Climate-related Financial Disclosures (G20 initiative) emphasized on the reinforcement of the same.

The main reason for under performance of emission trade is relatively low prices of carbon for a number of years. There are also huge variations in the pricing. In 2017, prices for a tonne of carbon dioxide ranged from below $1 in Mexico and Poland to $126 in Sweden. In most places prices remain less than $10 a tonne. The difficulty in valuation of the social costs is one of the major causes of such huge variation. There has been difficulty in assessing the baseline of emissions for different types of entities. It is largely believed that carbon prices need to rise to $20 per tonne to make the investments lucrative. Some opined to raise the prices above $40 by 2025.

Clean Development Mechanism

The Clean Development Mechanism (CDM) is one of the Flexible Mechanisms defined in Article 12 of the Kyoto Protocol. The purpose of the CDM is to promote clean development with sustainability through emission reduction projects in developing countries. The CDM is primarily a project based mechanism which is designed by developed/industrialised countries to promote projects in the developing countries that helps in reduction in emissions. The CDM is sometimes referred as production based mechanism as this project produces an 'emission cut' and it is subtracted against a hypothetical "baseline" of emissions (emission that would occur in absence of CDM project). In other words, the developing countries receive credit for doing these emission cuts.

The logic behind the implementation of these projects in developing countries only, is mainly the substantially lesser expenditure incurred on emission cuts in developing countries compared to developed countries. For example, in developing countries, weaker environmental regulation generally allows for easy implementation of any project compared to developed countries. Thus, it is widely thought that there is a greater potential for developing countries to reduce their emissions than developed countries.

The CDM, was intended to meet two objectives,

(i) to assist parties not to include the industrialized countries in contributing to the ultimate objective of UNFCCC, and (ii) to assist parties of industrialized countries in achieving compliance with their quantified emission limitation and reduction commitments (GHG emission caps).

The CDM made provisions for emissions reduction by generating a kind of emission currency called "Certified Emission Reduction units (CERs)" which may be traded in emissions trading schemes.

The CDM allows the industrialized countries to meet part of their emission reduction commitments by buying CERs units from CDM emission reduction projects in developing countries. The prior approval has to be taken before setting up the project on CER issues to ensure that the emission reductions are real and "additional". The project is supervised by the CDM Executive Board (CDM EB) under the guidance of the Conference of the Parties (COP/MOP) of the United Nations Framework Convention on Climate Change (UNFCCC). The CDM allows industrialized countries to buy CERs and to invest in emission reductions, where it is cheapest globally. Till date, more than a billion CER units have been issued under CDM after its implementation. Most of the CERs had been issued for projects based on destroying either HFC-23 or N_2O.

In 2008, Delhi Metro Rail Corporation Limited (DMRC) become the first Railway project to be awarded under the CDM by the United Nations. The registration enabled the DMRC to claim carbon credits. Delhi Metro has received Rs 9.5 crores through CDM projects i.e regenerative braking and modal shift projects (2004 and 2012) under the Japan Finance Carbon Ltd. The DMRC is presently making attempts to create new projects viz., installation of a solar power plant at the Metro stations, and get them registered under Gold Standard.

Since, this mechanism involves the participation of two or more countries, the host country is required to ensure that the registered projects contribute towards tangible development. Some of the activities are also prohibited under CDM, such as use of nuclear power and setting up infrastructure that lead to deforestation or any form of ecological damage. The framework has provisions that promote supplemental activities to domestic actions. This provision helps in preventing the industrialised countries from making unlimited profits through CDM.

Joint Implementation

Joint implementation (JI) is the third flexibility mechanisms articulated in the Kyoto Protocol. The JI is legally binding to the industrialized countries (only) to meet the assigned amount of emission reduction under Kyoto Protocol. Under Article 6, JI was proposed as an alternative to reduce emissions domestically. Any industrialized country can invest in an emission reduction project in any other industrialized country, where the cost of emission reduction is expected to be cheaper. Various projects may be taken up under the JI which may involve replacing a less efficient system with a more efficient system. The earned credits may be used towards their own emission reduction commitment goals domestically. Presently, Russia and Ukraine are leading maximum number of JI projects.

The process of receiving credit for JI projects is different from CDM projects. Emission reduction under JI projects are awarded credits called Emission Reduction Units (ERUs), which represents an emission reduction equivalent to one tCO_2e. The ERUs are calculated on the basis of host country's share of assigned emissions credits, known as Assigned Amount Units, or AAUs. Each industrialized countries has a set amount of AAUs, calculated on the basis of its 1990 greenhouse gas emission levels. The JI credits mandatorily come from a share of AAUs of the host country.

7.1.2 Ecosystem services

It is inevitable to regulate the resources or rather services that we receive from the environment or ecosystem in the changing climatic scenario. Earlier in this chapter, we came to know about how emissions are converted into commodities and traded for higher economical gains. The concept of ecosystem services is somewhat similar, wherein, the ecosystem functions and components are translated into commodities using economic or market based logics. In simple words, ecosystem services is the way to add economic value to the nature. The Millennium Ecosystem Assessment (MA) report 2005 defines the ecosystem services simply as "benefits obtained from ecosystems" and it is categorized into four types of services: (i) supporting services, (ii) provisioning services, (iii) regulating services, and (iv) cultural services.

Supporting services include nutrient cycling, primary production, soil formation, habitat provision and pollination. These services are essential for supply of basic needs (services), like food and water.

Provisioning services include the crops (food, fodder, fibre), water, raw materials (timber, skins, fuel, fodder, bio fertilizer), genetic resources (superior genes, health care), biogenic minerals, medicinal resources (pharmaceuticals, chemicals, and test and assay organisms), energy (hydropower, biofuels), etc. that can be directly utilized for the benefit of mankind.

Regulating services are the indirect services that the nature/ ecosystem provide to us. Carbon sequestration and climate regulation, prey populations regulation through predation, waste decomposition and detoxification by microbes, purification of water and air, biological pest and disease control are the major examples of those.

Cultural services include use of nature as inspiration for books, film, painting, folklore, national symbols, architect (aesthetic significance) etc. The cultural services hold recreational (ecological parks), educational, and/ or spiritual significance.

The Common International Classification of Ecosystem Services (CICES) is a classification scheme developed to accounting systems, in order to avoid double-counting of supporting services with others provisioning and regulating services.

How carbon sequestration qualifies for ecosystem services

Carbon sequestration is classified under the regulating services under the ecosystem services. After the industrial revolution the anthropogenic carbon emissions has increased at an average growth rate of about 2% per year. However, the atmosphere contains less than half of this fraction, rest of the carbon is removed by land and ocean sinks indicating a high self-regulating capacity. We still need to regulate the carbon cycle for more and more carbon sequestration, as we stand on the threshold of anticipated human induced climate disaster. However, the ecologists and economists view on carbon sequestration vary. The ecologist viewpoint (Daily 1997) focuses on the function and process that are directly or indirectly related to the ecosystem viz., effect of high CO_2 levels in atmosphere on beneficial soil microbes, while the economist considers the benefits of ecosystem services that are directly consumed or used (Popp et. al., 2011; Smith et. al., 2013b). Ecologically, different management activities can be explored for increasing its carbon sequestration potential of any system like soil, forest, agriculture. Various soil and plant management techniques can be used to enhance the photosynthetic efficiency for terrestrial C sequestration. In economic terms, ecosystem services are recognised and valuated for protection of ecosystems. The carbon market also could provide incentives for reducing greenhouse gas emissions. It can assign values to the amount of carbon saved or sequestered.

7.2 Climate change policies: Past and present

Brief history of climate change policy

The awareness to formulate policies related to climate change is a relatively recent phenomenon considering the history of climate change. The climate has been under a continuous change since evolution of agriculture and more so after industrial revolution. Through in the 19th century, great scientists like Fourier, Tyndale, and Arrhenius successfully identified the role of greenhouse gases and their potential to increase the atmospheric temperatures. It was in the late 20th century, late 1970s to be more precise; the World Meteorological Organization (WMO) recognized the consequences of the human induced GHGs emission. In 1988, International Panel on Climate Change (IPCC) was constituted by the WMO and the United Nations Environment Programme (UNEP) to make assessment and compile all the information regarding the developments in the area of climate change research. The IPCC was given the responsibility to publish the compiled report, so that international communities can make some decisions based on scientific evidences. Time to time, the IPCC has been holding the conferences, debates and processes related to the climate change policies. In 1990, the first assessment report was drafted and released for the United Nations Framework Convention on Climate Change (UNFCCC).

The UNFCCC established the basic principles for negotiations related to climate change mitigation options. In 1997, Conference of Parties (COP) meeting took place in Kyoto, Japan, as a result of the negotiations undertaken by UNFCCC. This conference's outcome is primarily known as Kyoto protocol. Then, in 1997, some of the developed counties like USA and Australia refused to ratify the Kyoto agreement. However, after the IPCC fourth assessment report in 2007, USA and Australian agreed to make some changes in their climate policies. Over the years, different countries, both developing and developed have agreed that global warming is real and happening. The developed countries have accepted that they have been historically responsible for the present state of CO_2 status in the atmosphere.

Timeline of milestones related to climate policies

1972: United Nations Conference on the Human Environment, Stockholm Conference; this is the UN's first major conference that is considered as the first milestone which focused on the international environmental issues.

1979: World Climate Conference, Geneva; First international meeting on climate change that established the World Climate Programme.

1987: The Montreal Protocol on Substances that Deplete the Ozone Layer, Montreal; the conference highlighted the damage of the ozone layer and its consequences. It advocated the restriction on use of chemicals that cause ozone layer depletion.

1988: Intergovernmental Panel on Climate Change, (IPCC) UNEP; the Intergovernmental Panel on Climate Change (IPCC) is a scientific and intergovernmental body under the United Nations. This body was constituted to publish scientific evidences related to climate change to provide current state of knowledge regarding climate change and its potential consequences.

1990: The first assessment report of IPCC concluded that the concentration of greenhouse gases is increasing at an alarming rate and it is mostly human induced activities that are responsible. The report emphasized that collective efforts need to be taken as the challenge is global and it cannot be tackle without full cooperation from all the countries.

1991: First meeting of the Intergovernmental Negotiating Committee; the meeting was part of the debates being conducted by IPCC on climate change policies. The Intergovernmental Negotiating Committee meeting played a decisive role in creation of the United Nations Framework Convention on Climate Change (UNFCCC).

1992: 'Convention on Climate Change', New York; the text of the UNFCCC is adopted at the 'United Nations Headquarters', New York. This was the key international treaty to reduce global warming and help to cope with the vagaries of climate change. It was for the first time the GHGs emissions reduction targets are set for industrialized countries.

1992: Earth summit (UN Conference on Environment and Development), Rio; All the agreed Governments are invited to sign the Convention on Climate Change at the Earth Summit in Rio.

1995: Conference of Parties (COP) 1, Berlin; It was collectively agreed that the commitments are largely inadequate to tackle the increasing emissions and therefore, stronger commitments from the developed countries are essential. It laid the base for the Kyoto Protocol.

1997: Introduction of Kyoto Protocol, Kyoto; The Kyoto Protocol is the world's first legally binding treaty related to reduction of GHG emissions.

2001: The Marrakesh accords (Seventh session of the conference of parties, COP7), Marrakesh; the Marrakesh Accords elaborates the implementation process of the Kyoto Protocol viz., funding sources, adaptation, strategies and development of new technologies. It emphasized the support needed by the developing countries in addressing the issues of climate change.

2005: The EU's Emissions Trading System is launched; World's first and largest emissions trading scheme, was launched.

2005: The 'Kyoto Protocol' comes into force.

2007: The IPCC's fourth assessment report was released; the summary of the climate change situation was published with the support of large number of references contributed by both scientists and governmental representatives. The report suggested that the 20th century global warming is very likely due to the increased anthropogenic activities.

2010: 'Cancun Agreements, Cancun'; in the 'Cancun Agreements' a comprehensive package was adopted by governments to assist the developing nations. The 'Green Climate Fund' was also established. The 'Cancún Agreements' which paved the way for a number of important arrangements, including the 'Green Climate Fund', the 'Technology Mechanism', the 'Cancún Adaptation Framework' and 'Forest Management Reference Levels'.

2011: Implementation of the 'UN Climate Change Convention' (7th session of the conference of parties, 'COP17'), Durban; in Durban, governments were clearly committed to a new universal climate change agreement by 2015 for the period beyond 2020, where all will play their part to the best of their ability and all will be able to reap the benefits together.

2012: 'Doha Amendment' to the 'Kyoto Protocol' (18th session of the conference of parties, 'COP18'), Doha; Governments were agreed to speedily work toward a universal climate change agreement by 2015 and to find the ways to scale up the efforts before 2020 beyond existing pledges to curb emissions. They also adopt the 'Doha Amendment' to the 'Kyoto Protocol', adding new emission reduction targets for participating countries for 2012-2020.

2013: 'Warsaw Agreement' (19th session of the conference of parties, 'COP19'), Warsaw; parties were agreed to a time plan for countries to table their intended contributions for the new global climate agreement and for ways to accelerate efforts before 2020. They set up a mechanism to address losses and damage caused by climate change in vulnerable developing countries. They also enhance the implementation of measures already agreed (for instance on climate finance) and transparency of reporting on emissions.

2015: This was the '20th session of the conference of parties', 'COP20', Lima; the 'COP20' requires all countries to describe their intended contributions for the 2015 agreement clearly, transparently and understandably.

2015: 'Paris Agreement', Paris; the 'Paris Agreement' is legally binding global climate deal. The agreement aimed to ensure controlled emissions and removal of GHGs in the 2nd half of this century. Furthermore, the agreement addresses the adaptation to climate change, financial and other support for developing countries, technology transfer and capacity building, as well as losses and damage.

2016: The 'Paris Agreement' enters into action.

2016: The 22nd session of the conference of parties, 'COP22', Marrakesh; parties pledge to move forward on the full implementation of the 'Paris Agreement' and welcomed the "extraordinary momentum on climate change worldwide". 'Marrakech Action Proclamation' for our climate and sustainable development. Countries at UN conference pledge to go ahead with implementation of 'Paris Agreement'.

2017: The 23rd session of the conference of parties, 'COP23', Bonn; the 'COP23' makes significant progress toward clear and comprehensive implementation guidelines of the Paris Agreement, which will make the agreement operational.

7.2.1 Efforts made by European Union

Apart from the efforts made by UN, individual countries have made commitments to tackle the climate change. Among all, Europe has been leaders in promoting action on climate change. At European level many packages of policy measures have been initiated through various programmes.

Before Kyoto Protocol, in 1991, the programme "Specific Actions for Vigorous Energy Efficiency" (SAVE) started to facilitate and promote the implementation of energy efficiency policies and programmes. ALTENER programme was introduced by the EU a Community-wide indicative target to promote renewable energies in 1993. The European Climate Change Programme (ECCP) was established in 2000. The second European Climate Change Programme (ECCP II) was launched in October 2005. It led to the introduction of the European Emissions Trading Scheme (ETS) with national caps for emissions from power and industry sectors in each member state, as well as proposals and communications on fields such as energy labelling, or the promotion of cogeneration and biofuels

The 2030 energy efficiency target will remain indicative at EU-level with targeted improvements of 27% over a baseline projection. Other instruments, such as the Ecodesign Directive and Eco-Labelling, or the CO_2 standards for cars may also be revised before 2020 to provide additional reductions. New policies for other sectors may be developed in the time-frame 2015-2020.

7.2.2 Asia Pacific Partnership (APP)

The 'Asia-Pacific Partnership on Clean Development and Climate', known informally as 'APP', is a non-treaty partnership, established by six nations namely, Australia, Japan, India, China, United States and South Korea in July 2005, which was formally launched in 2006. The partnership involved countries that account for about half of the world's population and more than half of the world's economy, energy use, and greenhouse gas emissions. In October 2007, In Canada the APP coiled the three major objectives among their partners. The objective were, (i) to expand markets for investments in cleaner, more efficient energy technologies, goods, and services in key sectors; (ii) to finance the initiatives for using green technologies in association with multilateral development banks; (iii) to work on areas of collaboration including energy efficiency, methane capture and re-use, clean coal, rural energy systems, advanced transportation, civilian nuclear power, geothermal, liquefied natural gas, bioenergy, agriculture, forestry, hydropower, wind power, solar energy and home construction.

7.2.3 Efforts made by India

Indian subcontinent has been contributing in CO_2 emission considerably. It contributes around 5% of total global CO_2 emissions. Further, the emission is projected to enhance by 70% during 2025-30. However, India's emissions are low if we compared with other developed countries. Moreover, India accounts for only 2% of cumulative energy-related emissions since 1850. But important to consider that on emission per capita basis, our country is 70% below to that of the world average and 93% below as compared to USA.

India's Policies on Climate Change

Indian government has been taken many initiatives to curtail down the GHGs emissions. For instance, Indian government released India's first National Action Plan on Climate Change (NAPCC) on June 2008. That plan has identified eight core "national missions" which was running through 2017. The goals of that plan were the emissions reduction intensity as per the commitment with the India's Copenhagen pledge.

National Action Plan on Climate Change (NAPCC): Eight Point Mission

The NAPCC primarily focused on eight point missions considering the climate-change related issues, renewable energy sector development and overall economic development of the country. The eight (8) missions of NAPCC are discussed briefly in coming sections.

National Solar Mission: The promotion and development of solar energy sector for power generation and other alternative uses. The major objective was to partially replace fossil-fuel energy sector with solar-energy sector. The initiatives includes establishment of solar research centres, strengthening of national manufacturing infrastructure and capacity, enhancement of international collaboration and off-course increment of government funding to solar-energy sector.

National Mission for Enhanced Energy Efficiency: In this mission the energy efficiencies of both small and big industries was tried to increase by issuing the trade energy-saving certificates to companies and facilitating the financing of public-private partnerships. In the demand side, training and management programs were encouraged in the agricultural sectors and also at urban-municipal corporation level. Reduction on taxes on energy-efficient appliances was also introduced.

National Mission on Sustainable Habitat: The urban planning by extending the existing ECBC code (Energy Conservation Building Code), strengthening the automotive fuel economy standards, and using optimum-pricing measures are initiated and executed in NAPCC mission. The purchase of efficient vehicles for public transportation was also encouraged. The mission also emphasizes the recycling and re-use of wastes.

National Water Mission: This mission was aimed to improve the water use efficiency by 20% through effective storage, proper pricing and conservation of this valuable resource in climate change perspective.

National Mission for Sustaining the Himalayan Ecosystem: This mission has very specific goal to prevent glacial-melting in the Himalayan region along with maintaining the biodiversity in the region.

Green India Mission: This was a very ambitious mission. It aims to reforestation / afforestation of about 6 million ha of degraded lands/ waste land and consequently expanding the forest area from 23 to 33% in India.

National Mission for Sustainable Agriculture: This mission aims to develop climate-resilient agriculture. The basic components were expansion of weather-based crop insurance mechanisms, weather forecasting and development of adaptive crop-production technologies including introduction of climate resilient crop varieties.

National Mission on Strategic Knowledge for Climate Change: A new Climate Science Research Fund was formulated for facilitating the programmes on climate modelling studies and development of better data inventories through international collaboration. The private sectors initiatives were also encouraged for developing the adaptation and mitigation strategies to cope up with climate change vagaries.

The five ongoing initiatives were also emphasised by NAPCC which are summarised below.

Power generation: The government has been supporting the research and development on IGCC (Integrated Gasification Combined Cycle) and supercritical technologies. And at the same time instructed to close-down the inefficient coal-fired power plants.

Renewable energy: The central and the state electricity regulatory commissions have to purchase a certain percentage of grid-based power from renewable sources (Electricity Act, 2003; National Tariff Policy, 2006)

Energy efficiency: The large energy-consuming industries have to undertake energy audits and the specific energy-labeling program (Energy Conservation Act, 2001).

Proposals for health sector: The critical assessment of increased burden of diseases due to climate change and provision of enhanced public health care services are main two aspect of this programme.

Implementation: The objectives, activities, strategies, timelines, monitoring and evaluation criteria has to be developed by concerned ministry and must be submitted to the Prime Minister's Council on Climate Change. The PM Council would be responsible for periodical reviewing and reporting on each mission's progress.

The recent Government policies are focussing on 5 major aspects:

i. **Renewable energy**

   The two major renewable energy-related policies were the 'Strategic Plan for New and Renewable Energy' (broad framework) and the "National Solar Mission', (sets a capacity targets for renewable). The "Solar Mission" was set the following targets for 2017: 27.3 GW wind; 4 GW solar; 5 GW bio-mass; and 5 GW other renewables. For 2022, these targets have been increased to: 20 GW solar; 7.3 GW biomass; and 6.6 GW for other renewables.

ii. **Solar**

   Our national government announced the development of infrastructure of solar-energy sector in its "National Solar Mission" to generate 100 GW-capacity-electricity by 2022. It is five-times more and over 30 times higher than the currently installed-infrastructure. The government also planned to invest in 25 solar parks, for increasing national solar-energy production capacity by tenfold.

iii. **Wind**

   In 'National Wind Energy Mission' the government recently announced plans to increase the wind-energy production to 50,000 to 60,000 MW by 2022. The Indian government is also planning to promote an offshore wind energy market.

iv. **Transportation**

   India announced a promising 'new vehicle fuel-economy standards' ('Indian Corporate Average Fuel Consumption standard') of 4.8 liters per 100 kilometers (49 MPG) by 2021-2022 (15% more efficient). Biofuel legislation was also set a target of 20% blending of ethanol and biodiesel in 2017.

v. **Smart Cities**

To address the challenges of urban development the Government has launched a programme to create 100 “smart cities” with better utilities, transport systems and energy networks. India’s ‘National Mission on Sustainable Habitat’ also includes initiatives such as the ‘Energy Conservation Building Code’, mandated for commercial buildings in eight states. The programme is aimed to support the waste management, recycling and improved urban planning.

7.3 Climate change mitigation and food security dilemma

In the past few decades, the developing countries like India, China, and other countries of South America, Asia and Africa have been witnessing tremendous economic growth. There is no hiding the fact that economic growth comes with industrial growth and, industrial growth comes at the expense of environment. The population pressure is high in most of these developing countries where the demands for supplies (agricultural and industrial) are meant for survival and sustenance. A small cut in the supplies means millions of unfed mouths which is unreasonable.

India is a paradox with high economic growth alongside high poverty and high economic inequalities. Presently, India supports 18% of world population with substantial percentage of the world’s poor. India’s food production has been able to keep pace with the growing population but it still houses about a fourth of the world’s malnourished. Any qualitative or quantitative improvement in the lives of people cannot be done without improved technologies. Here comes the importance of energy requirement. Almost all improved technologies require energy. In agriculture, huge amount of energy is required (for managing the crops, harvesting, processing, packaging, etc.). Not just agriculture, all other allied sectors (industry, animal husbandry etc.), are heavily dependent on energy. Presently, India’s energy demands are met by coal (>50%), oil (~30%), natural gas (~10%) and others (hydropower, nuclear energy etc.). Efforts are being made to use the energy from coal more efficiently. Several initiatives have been taken to increase the coal conversion efficiency. To reduce the dependence on coal, the NAPCC aims to derive 15% of India’s electricity from renewable sources by 2020. However, the renewable sources that are available till date are able to contribute only a minuscule amount of the total energy demand. Food production will be badly affected if energy demands are not fulfilled.

7.4 Way forward

There remain some uncertainties regarding the climate change starting from the relevance of data and its interpretation to the future projections. However, most of the scientists and policy makers believe that global warming is real and happening. They understand the magnitude of damage that can be caused if this remains unabated. But in practice, the humanity as a whole is not willing to make any sacrifices. The world's largest polluter, United States of America has refuted to ratify the Kyoto Protocol as well as Paris Agreement. So far, the climate mitigation conferences and agreements have not been able to bring down the GHG emissions. It has been observed that the GHG emissions reduce only on the occasions of slower economic growth (for example, 2008-09). So, it is quite evident that curbing the emissions is possible only with low economic growth. The developing countries like India are in a more vulnerable position which needs faster economic growth to sustain its development. The government has to priorities the economic development, more so in the developing countries because the lives of poor improve only when the economy grows. The economic developments cannot be endangered for the potential future climate impacts. It has been proved and mutually agreed that few developed countries are contributing more towards climate change compared to the others. Therefore, these countries must take responsibilities to curb the emission to slow down the process of climate change. Under the given scenario, reducing the energy consumption is suicidal for any government. Countries like India are well aware of the consequences of climate change but they need to keep the balance as they have to fulfill the demands of the present generation without affecting the prospects of future generations.

7.5 Probable questions

i. What are the roles of IPCC?
ii. Describe important climate change conference and their implications.
iii. What are the key issues which should be addressed by policy makers for developing sustainable GHGs mitigation measures?
iv. Describe the challenges and opportunities of implementing GHGs mitigation measures in agriculture both in developing and developed countries.

References

Popp A., J. P. Dietrich, H. Lotze-Campen, D. Klein, N. Bauer, M. Krause, T. Beringer, D. Gerten, and O. Edenhofer (2011). The economic potential of bioenergy for climate change mitigation with special attention given to implications for the land system. Environmental Research Letters 6, 34 – 44. doi: 10.1088/1748-9326 / 6 / 3 / 034017, ISSN: 1748-9326.

Smith P., H. Haberl, A. Popp, K. Erb, C. Lauk, R. Harper, F. N. Tubiello, A. De Siqueira Pinto, M. Jafari, S. Sohi, O. Masera, H. Böttcher, G. Berndes, M. Bustamante, H. Ahammad, H. Clark, H. Dong, E. A. Elsiddig, C. Mbow, N. H. Ravindranath, C. W. Rice, C. Robledo Abad, A. Romanovskaya, F. Sperling, M. Herrero, J. I. House, and S. Rose (2013b). How much landbased greenhouse gas mitigation can be achieved without compromising food security and environmental goals? Global Change Biology 19, 2285 – 2302. doi: 10.1111 / gcb.12160, ISSN: 1365-2486.

Glossary

Adaptation Capacity: It is the intrinsic adaptive ability of crops and animals to cope up with climate variability.

Aerenchyma: A soft plant tissue containing air spaces.

Aerosol: A colloidal suspension of particles dispersed in air or gas.

Anthropogenic activities: An effect or object resulting from human activity.

Carbon cycle: It is the combination of different chemical pathways through which carbon flows between the earth-atmosphere systems.

Carbon sequestration: It is the process of storing carbon either in soil, plant, geological rocks or ocean for a longer period of time

Clean Development Mechanism (CDM): It is to promote clean development with sustainability through emission reduction projects in developing countries.

Climate Change: Climate change is refers to considerable and relatively persistent changes in the global climate.

Climate index: Climate index referred to the estimated values used to elaborate the state and the changes in the climate systems.

Climate Resilient Agriculture (CRA): It means the ability of the agricultural system to bounce back to its original when subjecting to certain climatic-stresses.

Climate Smart Agriculture (CSA): It is an holistic and integrated approach which considers three basic component of development, techniques, policy and infrastructure (investment) to achieve sustainable development in agricultural for food security under climate change scenarios.

Climate system: It is a complex system representing the five major components, namely, atmosphere, biosphere, hydrosphere, cryosphere and lithosphere.

Climate: Climate is the long term (>30 years) state of weather prevailing in an particular region/ area/ state/ country/ continent.

Crop-diversification: Growing of different crops in a particular farming system considering the plant architecture, soil and weather condition, economics and demand.

Cryosphere: The cryosphere is the portion of the earth-system including snow, ice sheets and ice selves, glaciers, ice on lakes and rivers, icebergs and sea ice, as well as seasonally frozen ground and permafrost.

Decomposition: It is the microbial mediated process by which complex organic matters are broken down into simpler organic substances.

Denitrification: The process by which the reduction of nitrates to nitrous oxide/ di-nitrogen/ nitrogen oxide in the presences of denitrifiers taken place.

Detoxification: The process of removing toxic substances.

Diffusion: The intermingling of substances by the natural movement of their particles.

Ecosystem respiration: It is the respiration of all living organism in a ecosystems including plant, animal and microbes.

Emission trading: The fundamental concept of this is to remunerate the emissions saved or capped in any industrial/ farming/ stake holder/ activity and it was named as emission trade.

Facultative anaerobe: Basically an aerobic microorganism but effectively survive and grow in the absence of oxygen.

Fertigation: Fertilizer mixing with irrigation water.

Flame ionization detector: It is generally used for detection of hydrocarbons that generate ions when heated with H2-air flame.

Flux: It is the movement of substance in a specific direction (upward/ downward/lateral/ horizontal/vertical) per unit area per unit time.

Mitigation of GHGs emission: It is the way or means by which the emission of GHGs could be reduced.

Global Warming Potential: It is a measure of how much heat a greenhouse gas traps/radiated back in the atmosphere up to a specific time horizon, relative to carbon dioxide.

Global Warming: The global surfaces are getting warmer at a rate higher than it expected for a specified time scale

Greenhouse Effect: The process by which greenhouse gases in atmosphere absorb and radiated back the long-wave radiations in earth-atmosphere and warm the planet surface temperature.

Hydrosphere: The hydrosphere is the combined mass of water found on, under, and above the surface of a planet.

Infrared (IR) radiation: It is a region of the electromagnetic radiation spectrum where wavelengths range from about 700 nm to 1 mm.

Methanogenesis: It is a microbial mediated process by which carbonaceous compounds, preferably acetate and or CO2 is converted to methane.

Methanotrophy: It is the oxidation of methane to carbon dioxide and water conversion of methyl group to CO2 occurred and O2 acts as electron acceptors.

Mineralization: It is the conversion of organically bound elements to mineral form which many a time water soluble and available to plant. It could be both chemical and microbial mediated.

Nitrification: Microbial mediated oxidation of ammonia/ ammonium to nitrite followed by nitrite to nitrate.

Nitrogen use efficiency (agronomic)**:** The ratio between the amount of fertilizer N

Pacific Decadal Oscillation: It is a frequent pattern of climate variability occurring in higher latitudes (north of 200N).

Perennial: Plant that lives more than two seasons.

Photosynthesis: It is the most important biochemical process in which, the plants use the solar energy (light) and produce carbohydrate with the help of water and carbon dioxide.

Radiation: The emission of energy as electromagnetic waves

Radiative forcing: It refers to the net energy balance change in the earth-atmosphere system due to the imposed perturbation.

Renewable energy: Those energy sources that is naturally replenished

Respiration: Biological process by which organic molecules are broken down and chemical energy is produced and by the process carbon dioxide is released to the atmosphere.

Rhizosphere: The zone of soil (medium) actively influenced by root activities.

Root exudation: it refers to a suite of substances (organic acids, sugars, metabolites, etc.) in the rhizosphere that are secreted by the roots of living plants.

Soil amendment: It refers to any substances that added to soil for improving soil physical and or chemical or biological properties/ processes.

Soil quality: It is the ability of a specific kind of soil to function optimally (for a specific goal), within natural or managed ecosystem boundaries.

Zero tillage: The process/ practice of growing crops or pasture from year to year (season to season) without disturbing the soil.